# Contents

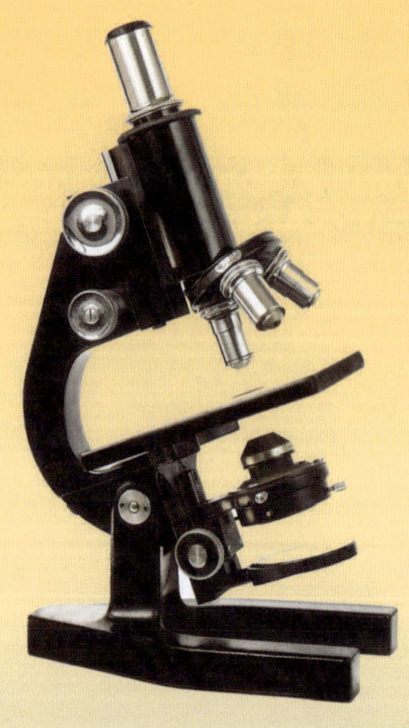

TARGETING SCIENCE YEAR 2 © PASCAL PRESS ISBN: 9781925726510

**Targeting Science Year 2**
Copyright © 2024 Pascal Press

ISBN: 9781925726510

Published by Pascal Press
PO Box 250
Glebe NSW 2037
www.pascalpress.com.au
contact@pascalpress.com.au

This edition of *Skill Sharpeners: Science* is published by arrangement with Evan-Moor Corporation, USA.

For sale in Australia and New Zealand.

Cover Design: Janice Bowles
Authors: Guadalupe Lopez, Lisa Vitarisi Mathews
Publisher: Lynn Dickinson
Editor Australian edition: Stella Tarakson
Typesetter: Stacey Grainger
Illustrator: Paul Lennon, *www.dreamstime.com.au*
Cover design: Janice Bowles

**Reproduction and communication for educational purposes**
The Australian *Copyright Act 1968* (the Act) allows a maximum of one chapter or 10% of the pages of this work, whichever is the greater, to be reproduced and/or communicated by any educational institution for its educational purposes provided that the educational institution (or that body that administers it) has given a remuneration notice to the Copyright Agency Limited under the Act.

For details of the Copyright Agency licence for educational institutions contact:
Copyright Agency
Level 12, 66 Goulburn Street
Sydney, NSW 2000
Tel: (02) 9394 7600
Email: memberservices@copyright.com.au

**Reproduction and communication for other purposes**
Except as permitted under the Act (for example a fair dealing for the purpose of study, research, criticism or review) no part of this book may be reproduced, stored in a retrieval system, communicated or transmitted in any form or by any means without prior written permission. All inquiries should be made to the publisher at the address above.

All material identified by AC | Australian CURRICULUM is material subject to copyright under the Australian *Copyright Act* 1968 (Cth) and is owned by the Australian Curriculum, Assessment and Reporting Authority 2017. For all Australian Curriculum material, this is an extract from the Australian Curriculum.

Disclaimer: ACARA neither endorses nor verifies the accuracy of the information provided and accepts no responsibility for incomplete or inaccurate information. In particular, ACARA does not endorse or verify that:
- The content descriptions are solely for a particular year and subject;
- All the content descriptions for that year and subject have been used; and
- The author's material aligns with the Australian Curriculum content descriptions for the relevant year and subject.

You can find the unaltered and most up-to-date version of the material at http://www.australiancurriculum.edu.au.

This material is reproduced with the permission of ACARA.

Printed in China by 1010 International Ltd.

## Introduction

Welcome to your Year 2 *Targeting Science* activity book! It is packed with interesting and exciting activities to help you understand and enjoy science at home or school.

*Targeting Science* has been written to support the Australian Primary Science Curriculum Version 9.0 and is divided between:

- Chemical Sciences
- Earth and Space Sciences
- Physical Sciences

The Science Understanding and Inquiry Skills elements (including Science as a Human Endeavour) are embedded.

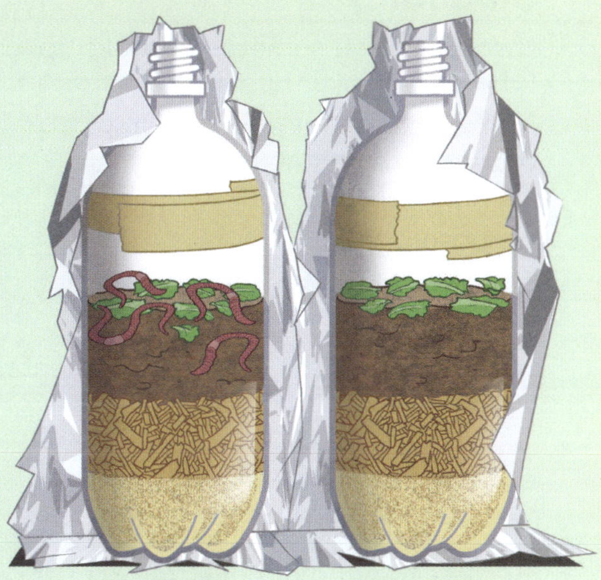

bottle with earthworms　　bottle without earthworms

Hands-on activities provide opportunities to bring the science concepts to life. The exercises develop process skills such as observing, collecting, recording and organising information and making models of scientific events that happen in the natural world. All activities use easily sourced, inexpensive items.

Topics are introduced with explanations plus images to provide background information. Some lessons include a QR code where you can access a video for further explanations. You will be challenged to match, sort, label, sequence, analyse and answer questions with regular vocabulary practice puzzles throughout the book to help familiarise you with scientific terms. See the Glossary on the next page for some terms you may not already know. Answers are included at the end of the book.

## FREE Teaching Guide.

This QR code links to a downloadable PDF of a Teaching Guide to support the material in this student workbook. The guide contains:

- Teaching plans and checklists.
- Graphic organisers.
- Material request forms (to send home for parents).
- Background information about each major topic as well as extension activities.

**Use this QR code to access the FREE Teaching Guide.**

Front cover – have you looked at the front cover? It depicts what life was like before a significant scientific discovery that led to the development of new technology. What is the difference between Science and Technology? The following quote sums it up nicely.

*Science is the process of acquiring knowledge of natural phenomenon along with various reasons. Technology is the application of Science to the solution of problems. (Sismondo, 2018)*

# Glossary

- **absorb** – when something 'soaks up' energy instead of reflecting it or letting it pass through.
- **constellations** – a group of stars that seem to form a design in the night sky.
- **designed** – made or done intentionally or for a special purpose.
- **echo** – a sound or sounds caused by the reflection of sound waves from a surface back to the listener.
- **echolocation** – the location of objects by reflected sound, used by dolphins and bats.
- **engineer** – a person who designs, builds and maintains engines, machines and structures.
- **moon** – a natural satellite that orbits a planet.
- **muffle** – to wrap or cover to prevent sound.
- **phases** – distinct stages of development. Moon phases are variations in the illumination and apparent shape of the moon.

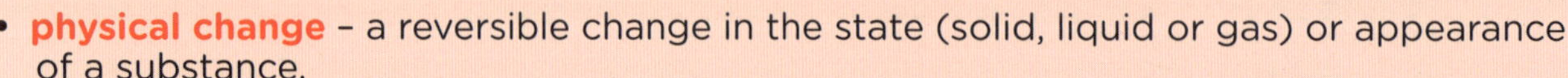

- **physical change** – a reversible change in the state (solid, liquid or gas) or appearance of a substance.
- **pitch** – how high or low a sound is.
- **properties** – attributes of an object or material, usually used to describe attributes common to a group.
- **scientist** – someone who uses scientific methods to answer questions about the universe.
- **stars** – massive self-luminous celestial bodies of gas that shine by radiation derived from their internal energy sources.
- **sun** – the star in the centre of the Solar System.
- **vibrations** – to-and-fro movements that produce sound.

**SAFETY**
**All of the investigations in this book are designed for kids to do safely in the home or classroom. However, adult supervision is recommended.**


TARGETING SCIENCE YEAR 2 © PASCAL PRESS ISBN: 9781925726510

# Year 2 Australian Science Curriculum Correlations

| ACARA Code | Content description | Strand | Sub strand | Pages |
|---|---|---|---|---|
| AC9S2U01 | Recognise Earth is a planet in the solar system and identify patterns in the changing position of the sun, moon, planets and stars in the sky | Science Understanding | Earth and Space Sciences | 2-35 |
| AC9S2U02 | Explore different actions to make sounds and how to make a variety of sounds, and recognise that sound energy causes objects to vibrate | Science Understanding | Physical Sciences | 36-75 |
| AC9S2U03 | Recognise that materials can be changed Describe how people use science in their daily lives, including using patterns to make scientific predictions | Science Understanding | Chemical Sciences | 76-117 |
| AC9S2H01 | Recognise that materials can be changed physically without changing their material composition and explore the effect of different actions on materials including bending, twisting, stretching and breaking into smaller pieces | Science as Human Endeavour | Use and Influence of Science | 33, 54, 56, 57, 73, 76, 78, 86, 88, 96, 116 |
| AC9S2I01 | Pose questions to explore observed simple patterns and relationships and make predictions based on experiences | Science Inquiry | Questioning and Predicting | 9, 42, 43, 54, 65 73, 77 |
| AC9S2I02 | Suggest and follow safe procedures to investigate questions and test predictions | Science Inquiry | Planning and Conducting | 10, 23, 25, 43, 54, 55, 73, 77 103, 104, 109 |
| AC9S2I03 | Make and record observations, including informal measurements, using digital tools as appropriate | Science Inquiry | Planning and Conducting | 9, 42, 43, 54, 55, 66, 77 |
| AC9S2I04 | Sort and order data and information and represent patterns, including with provided tables and visual or physical models | Science Inquiry | Processing, Modelling and Analysing | 14, 43, 54, 91 |
| AC9S2I05 | Compare observations with predictions and others' observations, consider if investigations are fair and identify further questions with guidance | Science Inquiry | Evaluating | 9, 53, 55, 56, 77, 102 |
| AC9S2I06 | Write and create texts to communicate observations, findings and ideas, using everyday and scientific vocabulary | Science Inquiry | Communicating | 8, 42, 45, 53, 67, 74, 87, 94, 96, 97, 102, 109 |

# The Earth From Space

These are all photos of the Earth taken from space. The light you see in the distance is the sun. The sun is much bigger than the Earth, but it is a VERY long way away, so it looks smaller from here.

What shape is the Earth?

What do you think the green parts you can see are?

What do you think the blue parts you can see are?

What do you think the white parts you can see are?

Is there more green or blue?

In which picture can you see the sun starting to appear over the horizon?

Can you see the lights on at night in photo 'D'? Why aren't they spread out evenly all over the land, do you think?

How do you think they took these photos of the Earth?

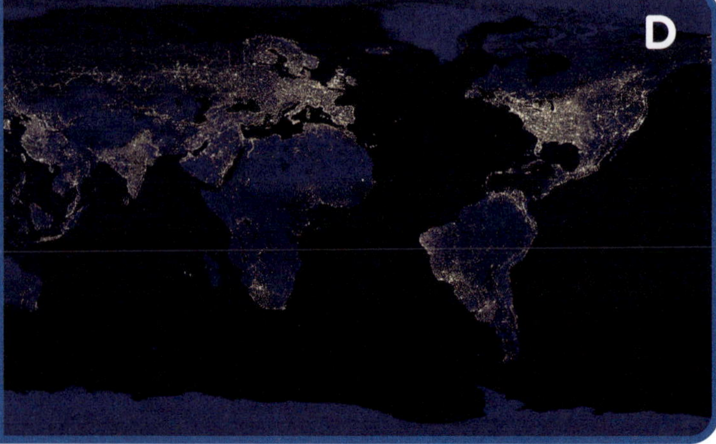


Solar System



TARGETING SCIENCE YEAR 2 © PASCAL PRESS ISBN: 9781925726510

This is a picture of the sun and the planets in our solar system. The sun is so big, you can't fit the whole thing into the picture, so only a part of it is shown.

Can you see tiny little Earth? It looks very small compared to the giant gas planets Jupiter, Saturn, Uranus and Neptune, doesn't it?

Did you know that you could fit 1.3 million Earths into the sun? It's huge! But when you are on Earth, it feels pretty big too, doesn't it?

The four planets closest to the sun are Mercury, Venus, Earth and Mars. They are made of rock.

The planets aren't really in a line like this. They are all travelling (or orbiting) around the sun, more like in this diagram. It's really hard for people to show just how big and far apart the planets are in the solar system. This picture shows how they orbit, but it's not as good as the top picture for showing the actual sizes, and it still doesn't show you just how far apart they are from each other.

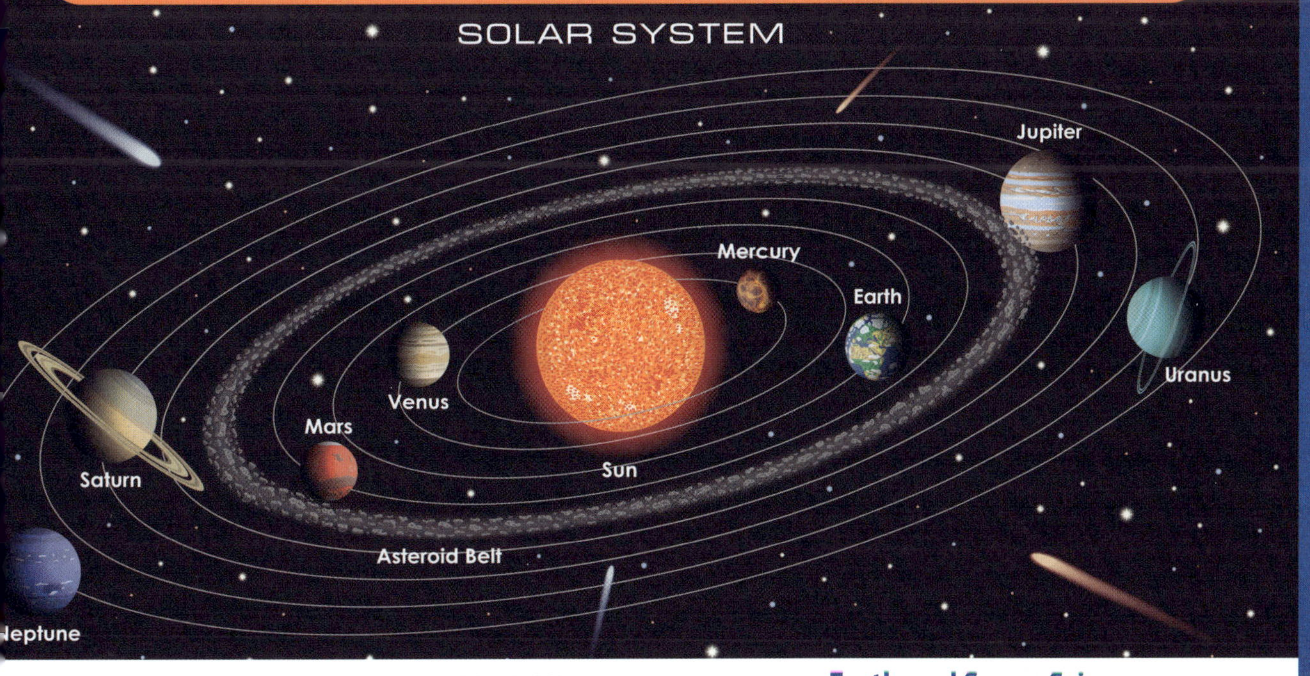

# Earth's Sun

Look at the picture. Read the word.
Write the word in the sentence.

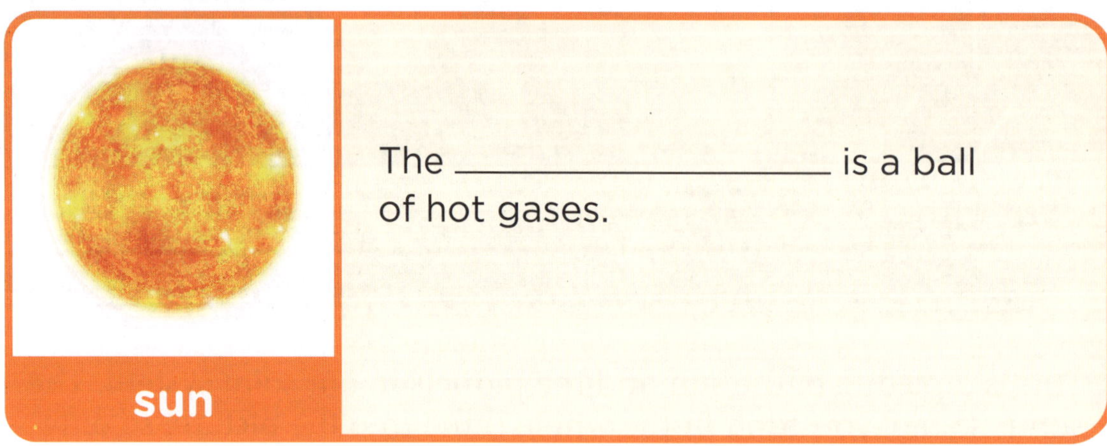


The ______________________ is a ball of hot gases.

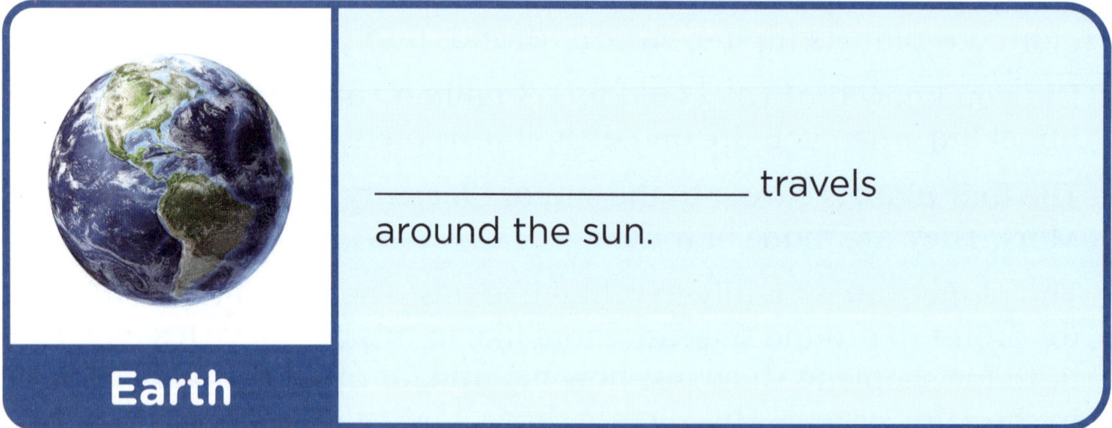


______________________ travels around the sun.

The sun helps ______________________ grow.

 TARGETING SCIENCE YEAR 2 © PASCAL PRESS ISBN: 9781925726510

The **sun** is an object in the sky. The sun is a ball of hot gases. It is the star that is closest to Earth. It is the star we see in the daytime.

The sun is much bigger than **Earth**. The sun looks small because it is so far away. Earth travels around the sun. It takes one year for Earth to travel all the way around the sun.

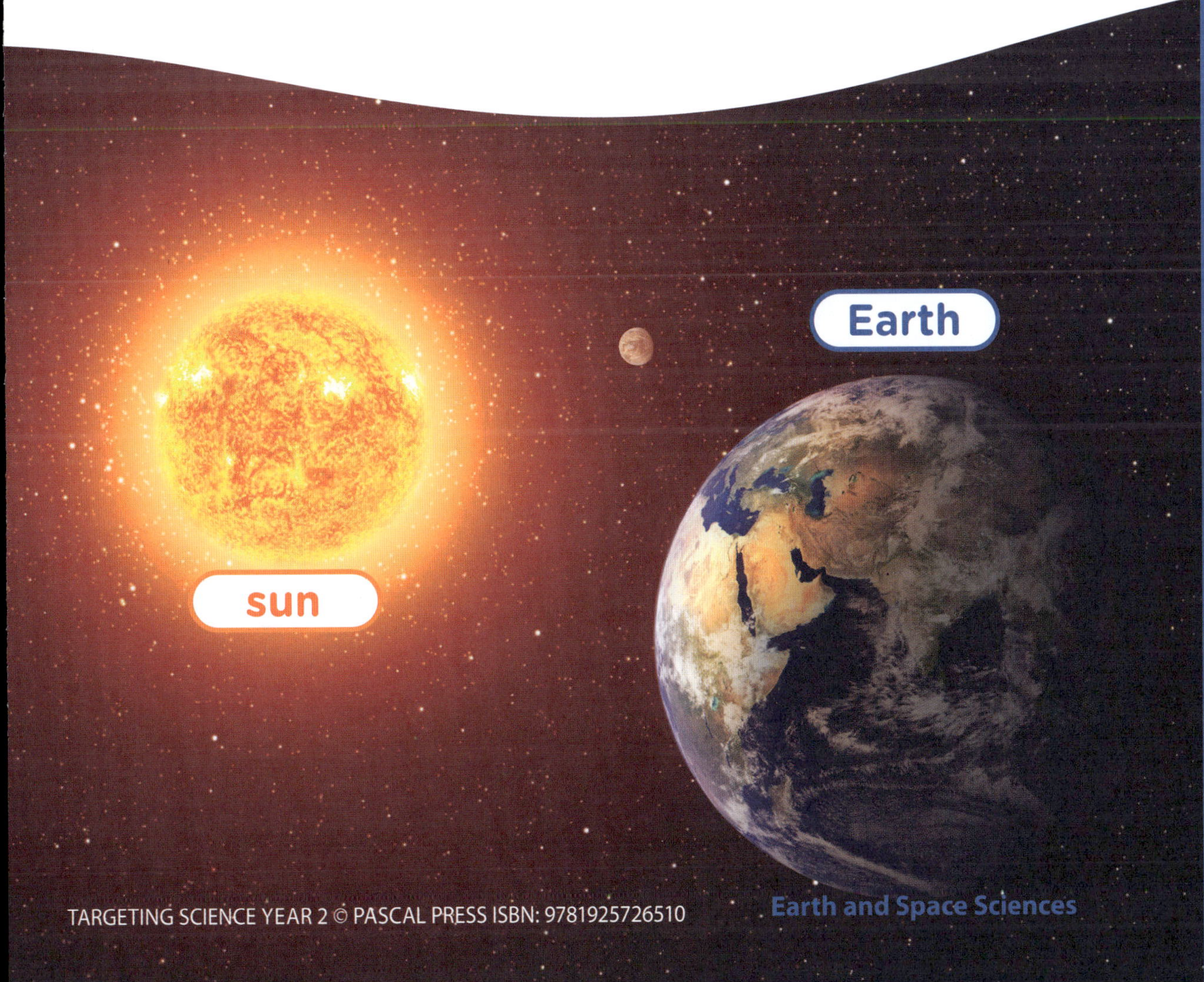

# We Need the Sun

People need the sun. It gives us heat to stay warm. It gives us light to see.

**Plants** need the sun, too. The sun helps plants make their food. Plants make oxygen for people and animals to breathe.

Without the sun, people, plants,
and animals
could not live on Earth.

TARGETING SCIENCE YEAR 2 © PASCAL PRESS ISBN: 9781925726510

Read the clue. Write the word.

1. The star closest to Earth

___ ___ ___
1

2. The sun gives off this so you can see.

___ ___ ___ ___ ___
2 3

3. The time it takes for Earth to travel all the way around the sun

___ ___ ___ ___

4. The sun gives us this to stay warm.

___ ___ ___ ___
4 5

Write the numbered letters to solve the puzzle.

**Science Puzzle**

We cannot see the sun at

___ ___ ___ ___ ___.
1 2 3 4 5

# The Sun

Look at this picture. Write about how the sun helps people, plants, and animals on Earth.

TARGETING SCIENCE YEAR 2 © PASCAL PRESS ISBN: 9781925726510

See how the light from the sun can cause changes.

## What You Need

- tape
- foam stickers or shapes
- black cardboard

## What You Do

1. Arrange the shapes on the card-board.
2. Tape the cardboard to a sunny window, facing out.
3. What do you think might happen? Why?
4. After a week, take down the cardboard and remove the shapes. Look at the cardboard and answer the question.

What did the sun do to the cardboard?

# Blended Suncatcher

Catch the sun with a disk that hangs in a window.

## What You Need

- glitter beads in see-through colours
- disposable pie pan
- parchment paper

## What You Do

1. Cut a circle of parchment paper to fit the bottom of the pie pan. Line the pan with the paper.
2. Ask an adult to preheat the oven to 190°C.
3. Arrange the beads on the parchment paper so that they look like a sun.
4. Place the pan in the oven for about 20 minutes or until the beads are melted.
5. Take the pan of melted beads out of the oven (an adult's job). Let them cool.
6. Pop the disk out of the pan. Drill a hole at the top (an adult's job).
7. String yarn through the hole and hang the sun in a window. Watch your bead sun catch the real sun!

TARGETING SCIENCE YEAR 2 © PASCAL PRESS ISBN: 9781925726510

Write to tell what you learned about the sun.

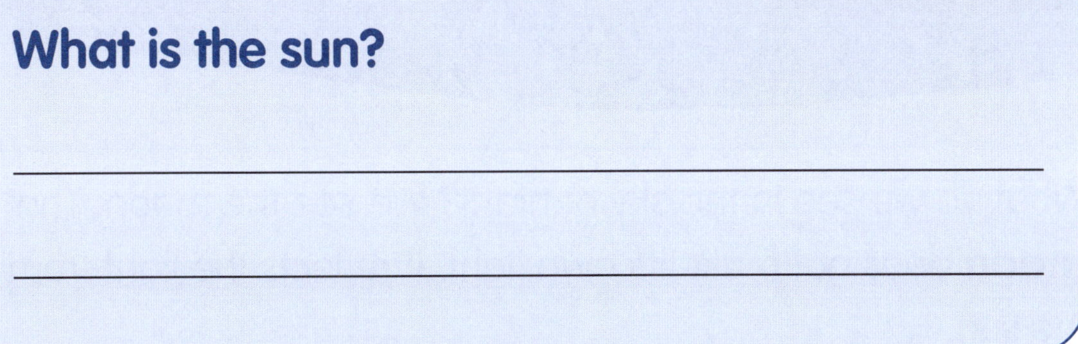

## What is the sun?

______________________________________________

______________________________________________

## Why does the sun look small?

______________________________________________

______________________________________________

## How does the sun help people, plants, and animals?

______________________________________________

______________________________________________



TARGETING SCIENCE YEAR 2 © PASCAL PRESS ISBN: 9781925726510

# We See the Moon

https://clickv.ie/w/szgx

Use this QR code to access a video on this topic.

new moon

crescent moon

quarter moon

full moon

What do we see in the sky at night? We see the moon. The **moon** does not make its own light. It reflects the light from the sun.

The moon is always round, but we do not see all of it every night. As the moon travels around Earth, it looks different to us. It starts out as the new moon. As the moon moves, its shape looks like a **crescent**. By the time the moon travels all the way around Earth, it looks like a round, full moon. The changes in the shape of the moon that we see are called **phases**.



TARGETING SCIENCE YEAR 2 © PASCAL PRESS ISBN: 9781925726510

Write the name of each phase of the moon.

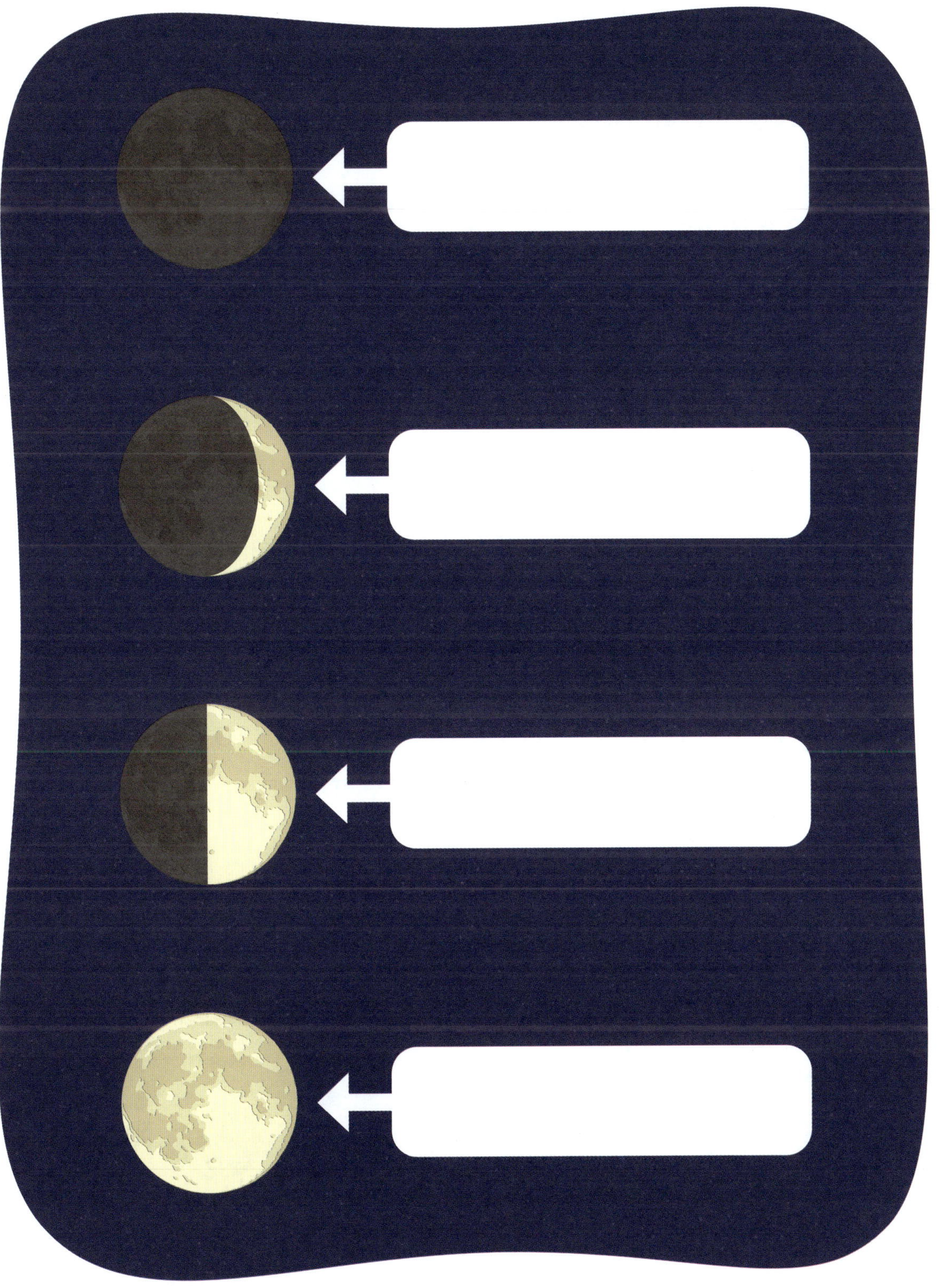

# Biscuit Moon

Show the phases of the moon using cream-filled biscuits.

## What You Need

- **8 cream-filled biscuits**
- **spoon**
- **paper plate**

## What You Do

1. Twist open the biscuits. Use only the sides that have filling. Set aside the other sides, except for one, which will be used as the new moon.
2. Use the spoon to scrape off the filling so that each biscuit looks like a phase of the moon.
3. Arrange the biscuits on a paper plate. Label each biscuit with the name of a phase: **new moon**, **crescent moon**, **quarter moon**, and **full moon**.

Space



TARGETING SCIENCE YEAR 2 © PASCAL PRESS ISBN: 9781925726510

The sun shines on the moon in different ways.

The part of the moon we can see changes.

We call these changes the phases of the moon.

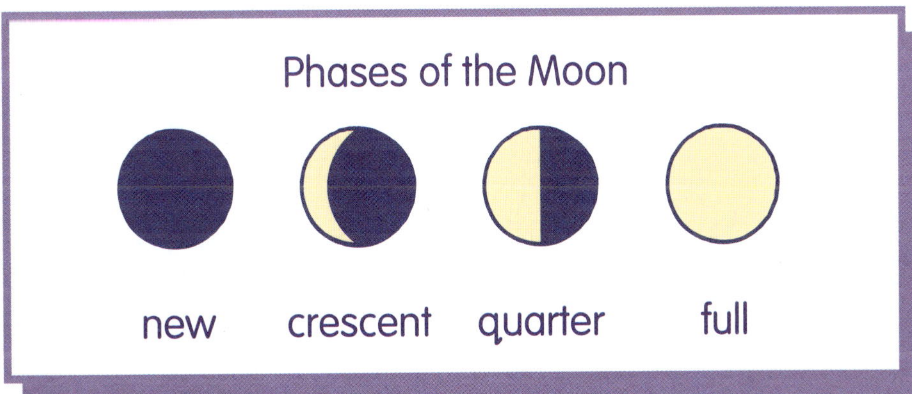




TARGETING SCIENCE YEAR 2 © PASCAL PRESS ISBN: 9781925726510 

# Moon Phases

Draw a line to match.

**new**

cannot see the moon

**quarter**

can see half of the moon

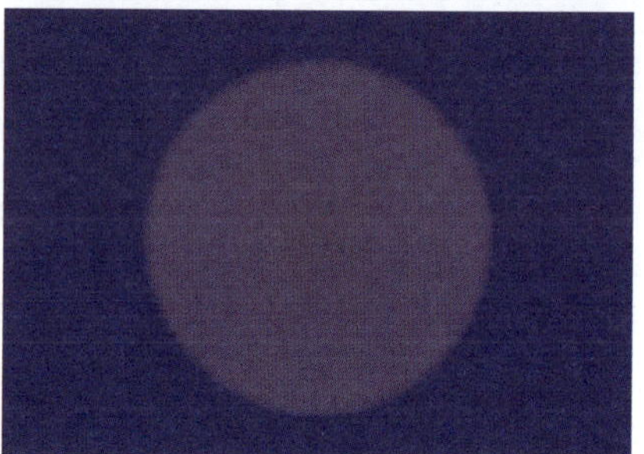

**full**

can see all of the moon

**crescent**

can see a small part of the moon

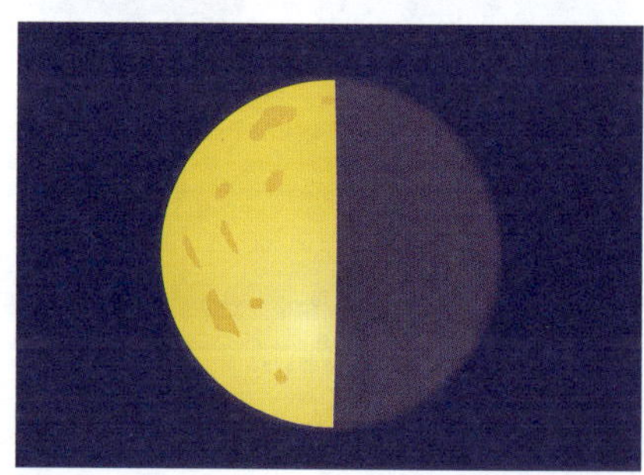

Make pictures that show the phases of the moon!

## What You Need

- 3 sheets of white paper
- 3 sheets of black paper
- 4 sheets of coloured cardboard paper
- a bowl or round traceable object
- crayons
- scissors
- glue

## What You Do

1. Set a bowl on a white sheet of paper. Trace around it. Cut it out. Do the same thing with each sheet of white paper.
2. Repeat Step 1 with the three sheets of black paper.

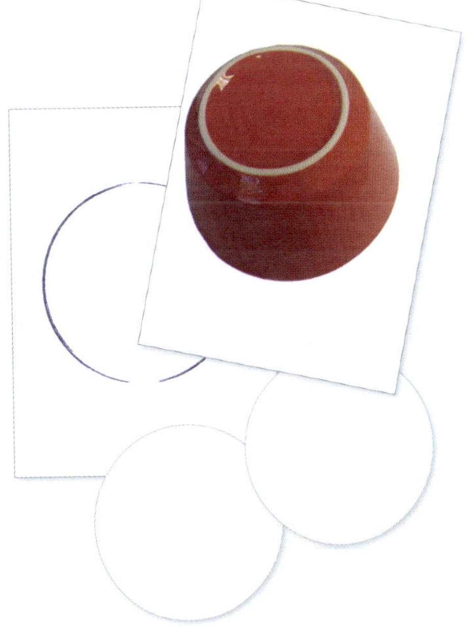

## Moon Art

**3** **Make a new moon:** Glue a black circle to a piece of cardboard. Label it “new moon”.

**4** **Make a crescent moon:** Cut a crescent shape out of a white circle. Glue it to a black circle. Glue it to the cardboard. Label it “crescent moon”.

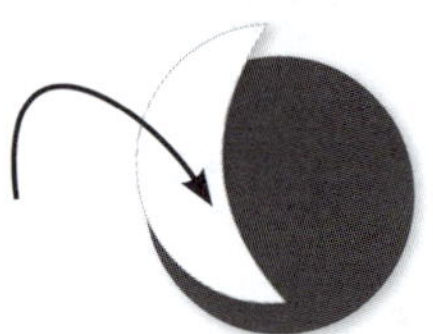

**5** **Make a quarter moon:** Cut one black circle in half. Glue one half of it onto one full white circle. Glue it to the cardboard. Label it “quarter moon”.

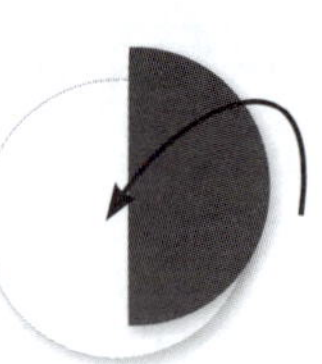

**6** **Make a full moon:** Glue a white circle to a sheet of cardboard. Label it “full moon”.

**7** **Share it:** Put your pictures in order and tell someone about the phases of the moon.

Step 3 Step 4 Step 5 Step 6


TARGETING SCIENCE YEAR 2 © PASCAL PRESS ISBN: 9781925726510

# The Changing Moon

Have you ever noticed that when you see the moon in the sky at night, it isn't always the same? Have you ever wondered why?

The moon's 'shape' is something we can predict. We can guess the way it will look ahead of time. The moon doesn't ever actually change its shape, it just has different parts that are lit up by the sun, and parts that are in shadow. It takes about a month for the moon to go through all of these shapes, or phases, and then they start all over again

So why does this happen?

The Earth is travelling or orbiting around the sun. The moon is orbiting around the Earth. Depending on where the Moon is compared to the Earth, different parts of it are blocked from the sun's light, making a shadow across it.

TARGETING SCIENCE YEAR 2 © PASCAL PRESS ISBN: 9781925726510

# Different pictures of the Moon

Every month, the moon goes through its phases, and follows the same pattern. Every now and then though, we see the moon look completely different in one of those phases.

| Full Moon | Blue Moon |
|---|---|
| We see the moon look like this every month. It is when the sun is shining directly on the moon without a shadow. | The moon can look blue sometimes due to smoke or dust being in the air. |

| Blood Moon | Super Moon |
|---|---|
| When the moon is completely in the Earth's shadow (a lunar eclipse) it can sometimes look red. | A Super Moon happens when the moon is full and much closer to the Earth in its orbit. |





TARGETING SCIENCE YEAR 2 © PASCAL PRESS ISBN: 9781925726510

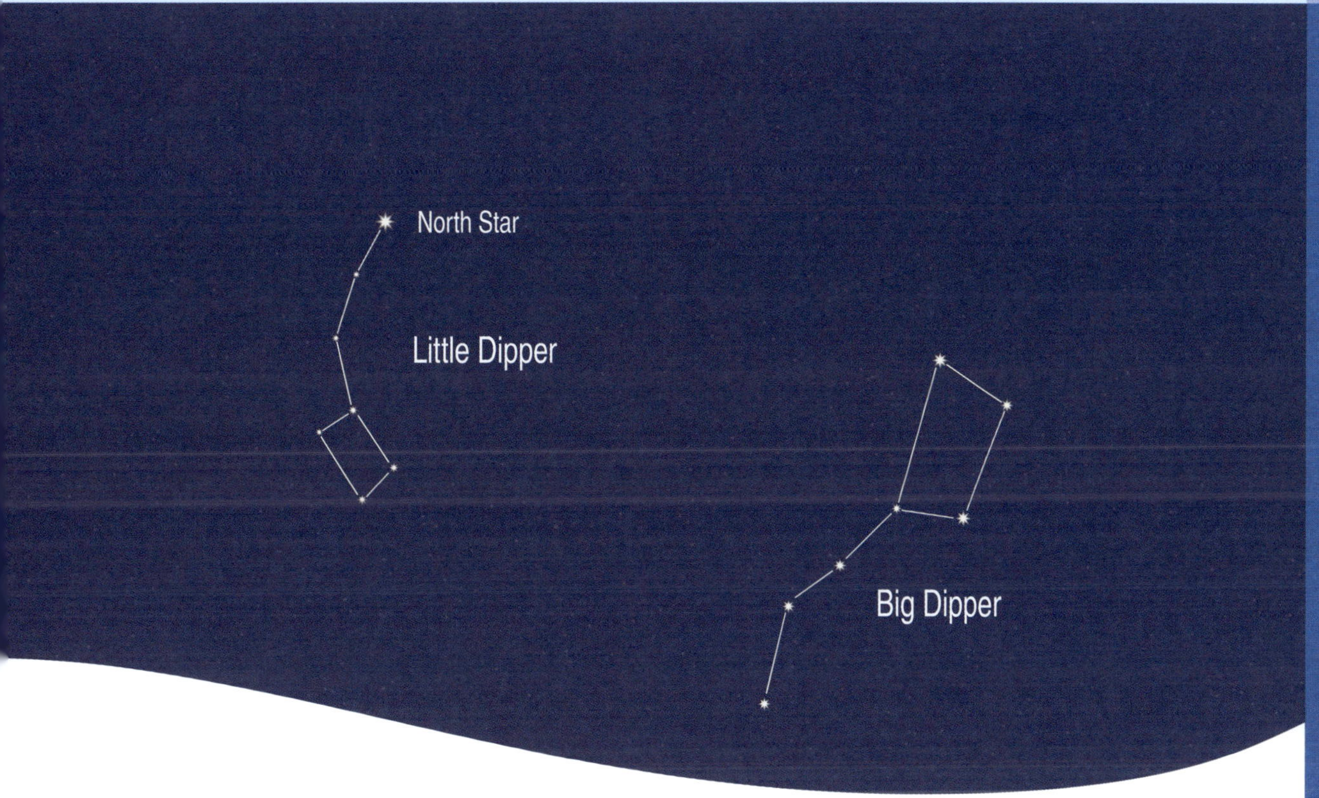


A star is a ball of hot gases. **Stars** make light and heat. Stars are so far away from Earth that we cannot feel their heat. The stars just look like tiny dots of light.

You can see about 2,000 stars with your eyes. Long ago, people gave names to groups of stars. These groups of stars are called **constellations**.

# Word Play

Read the clue. Write the word.

1. Big, round moon

____ ____ ____ ____   ____ ____ ____ ____
(letter 3 = 6)

2. Look like tiny dots of light

____ ____ ____ ____ ____
(letter 3 = 7, letter 5 = 3)

3. The planet you are on

____ ____ ____ ____ ____
(letter 1 = 5, letter 4 = 4)

4. The moon when it is shaped like a banana

____ ____ ____ ____ ____ ____ ____ ____   ____ ____ ____ ____
(letter 1 = 1, letter 7 = 2)

**Write the numbered letters to solve the puzzle.**

## Science Puzzle

The Big Dipper is a

____o____ ____ ____ ____ ____ ____ ____ ____io____.
1 2 3 4 5 6 6 7 4 2


TARGETING SCIENCE YEAR 2 © PASCAL PRESS ISBN: 9781925726510

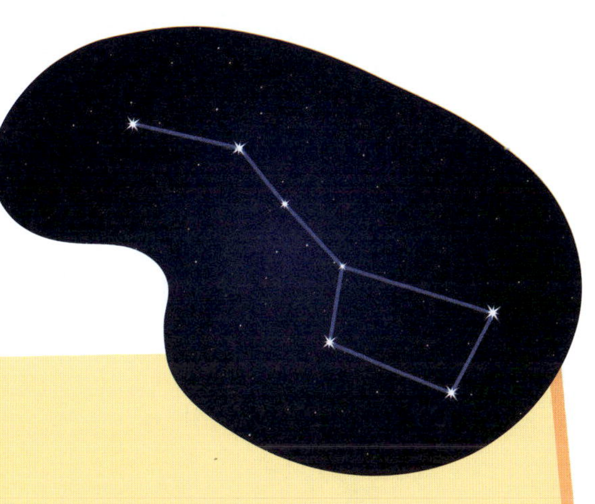

See the Big Dipper constellation on your wall!

## What You Need

- picture of the Big Dipper
- black paper
- clear jar
- glow stick
- sharpened pencil
- thumbtack
- chalk or white crayon

## What You Do

1. Cut the black paper so that it will fit like a cylinder inside the jar.
2. Using chalk or a white crayon, draw the Big Dipper on the black paper.
3. Poke holes in each star of the constellation. Use a pencil for big stars and a thumbtack for tiny stars.
4. Bend the black paper into a cylinder shape. Tape it closed. Place it inside the jar.
5. Put a glow stick inside the cylinder. Turn off the lights, and the constellation will appear on your wall!



TARGETING SCIENCE YEAR 2 © PASCAL PRESS ISBN: 9781925726510

# Moon and Stars

Draw a line to match.

| | |
|---|---|
| crescent moon | ball of hot gases |
| full moon | slice of moon |
| star | pattern of stars |
| constellation | changes in the moon |
| phases | big, round moon |


TARGETING SCIENCE YEAR 2 © PASCAL PRESS ISBN: 9781925726510

Draw a line to match.

star

full moon

sun

crescent moon

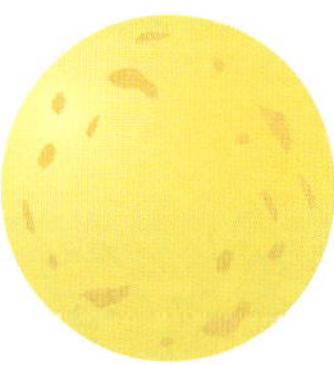

Finish the sentence.

The moon gets light from the ______________________.

# Night Sky Words

Unscramble the words about the night sky.
Write the words in the boxes.

**Word Box**

| star | light | moon |
|---|---|---|
| sun | phases | shine |

esshap

gilth

hisen

nus

omon

rast



TARGETING SCIENCE YEAR 2 © PASCAL PRESS ISBN: 9781925726510

Finish each sentence.

moon light sun

1. Stars make their own ______________________.

2. Light from the ______________________ makes the moon shine bright.

3. The part of the ______________________ we can see changes.

Draw your favourite phase of the moon.

# Moon and Stars

Look at the picture. Read the word.
Write the word in the sentence.

**moon**

The ________________ reflects light from the sun.

**crescent**

A ________________ shows only part of the moon.

**star**

A ________________ is a ball of hot gases.

**constellation**

A ________________ is a pattern of stars.


TARGETING SCIENCE YEAR 2 © PASCAL PRESS ISBN: 9781925726510

The weather changes each season.

Spring can be warm and rainy.

Summer can be hot and sunny.

Autumn can be cool and windy.

Winter can be cold and snowy.

Then the pattern starts all over with spring!

# Season Match

Draw a line to match.

winter

spring

summer

autumn

Finish the sentence.

A season is a time of ______________________.


TARGETING SCIENCE YEAR 2 © PASCAL PRESS ISBN: 9781925726510

Unscramble the words about the seasons.
Write the words in the boxes.

**Word Box**

| summer | autumn | spring | winter |
|---|---|---|---|
| season | warm | cold | hot |

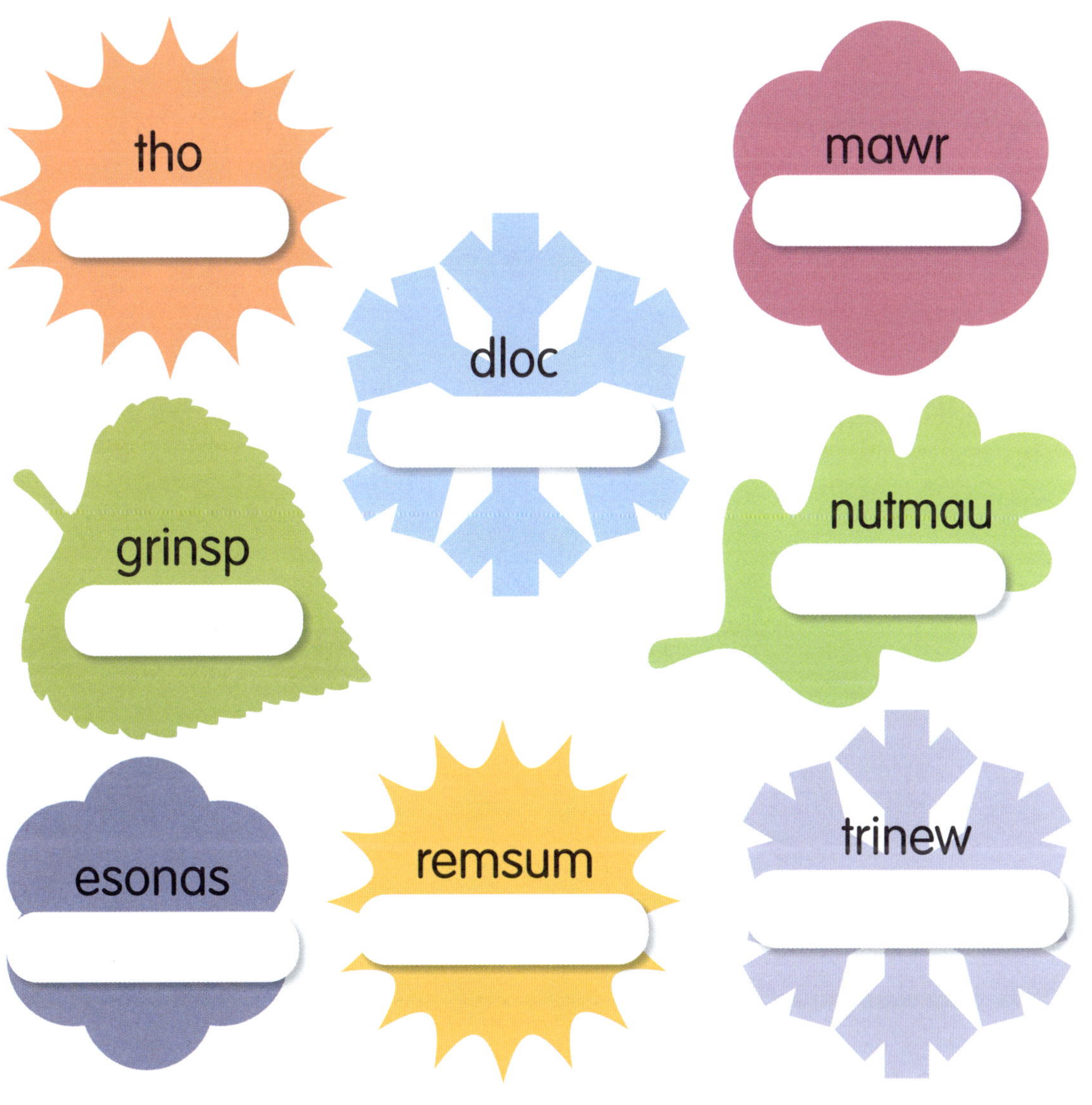




TARGETING SCIENCE YEAR 2 © PASCAL PRESS ISBN: 9781925726510 

Finish each sentence.

| weather | year | seasons |
|---|---|---|

1. Spring, summer, winter, and autumn are ______________

2. A season is a time of______________________.

3. The ______________________ changes each season.

Draw a picture of a tree in the autumn with falling leaves.

TARGETING SCIENCE YEAR 2 © PASCAL PRESS ISBN: 9781925726510

# Fingerpaint Trees

Use your hands to make trees for four seasons.

- white paper
- coloured cardboard
- paints
- scissors
- glue
- paper plates

## What You Do

1. Cut 2 sheets of white paper in half to make four sheets.
2. Pour brown paint onto a paper plate. Dip your hand and your wrist into the paint and press your hand and wrist onto one sheet of white paper to make the tree trunk and branches.
3. Do the same thing on each sheet of white paper. Let your "trees" dry.
4. Pour small amounts of coloured paints onto paper plates. Dip the tip of your finger into one colour at a time and fingerpaint leaves on the trees for each season.
5. After your trees are dry, glue them to the cardboard in order of the seasons: spring, summer, autumn, winter.
6. Write about your painting.

# The Changing Positions of the Sun, Stars and Moon: A First Nations Perspective

First Nations people look at the sun, stars and moon to predict when the seasons are changing, and how the plants and animals will behave.

They watch the patterns of the stars and how they move across the sky each year so that they can find their way across the land, sea and waterways.

They pass their knowledge on to younger people through their songs, stories and dance.

What story do you think these First Nations men could be telling through their dance?

Could you come up with a dance that you could do to tell a story?

TARGETING SCIENCE YEAR 2 © PASCAL PRESS ISBN: 9781925726510

Look at these amazing pictures. Can you work out what they are and how they were taken?

https://clickv.ie/w/wtgx

Use this QR code to access a video on this topic.

They are called 'time-lapse' photography. They are taken by leaving a camera in the same place for a really long time, recording what is happening.

The circles you see are images of the same stars being photographed as they move across the sky. They are making circles, because the Earth is spinning. It's a bit like if you stood in your bedroom and looked up at the light and spun around with paint on your feet. You'd make a lot of circles on the carpet, but your mum wouldn't like it!

https://clickv.ie/w/jygx

Use this QR code to access a video on this topic.

When light cannot move through something, you may see a **shadow**.

A shadow is a dark spot.



TARGETING SCIENCE YEAR 2 © PASCAL PRESS ISBN: 9781925726510

Draw a line to match.

shadow

glass

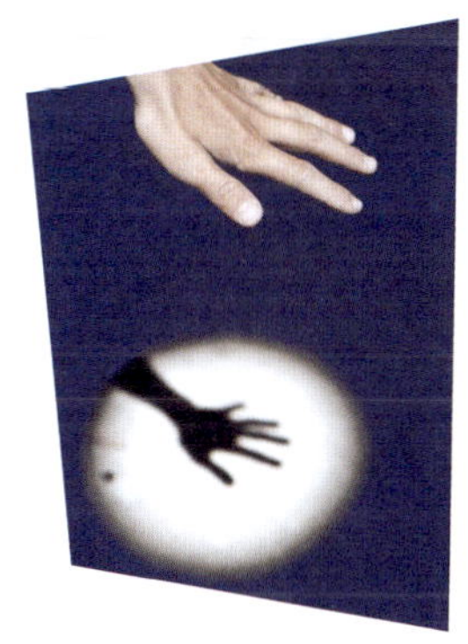

light

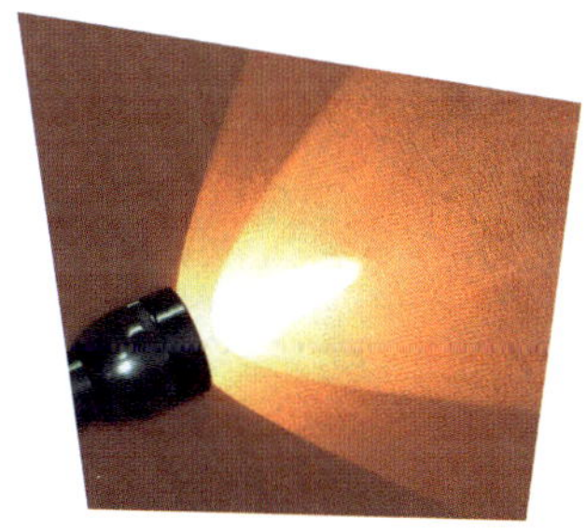

Finish the sentence.

When light cannot move through something,

it makes a ____________________.

# What Made the Shadow?

Draw a line to match.

TARGETING SCIENCE YEAR 2 © PASCAL PRESS ISBN: 9781925726510

Read the words in the word box.
Find the hidden words. Circle them.

**Word Box**

| shadow | light | glass | dark |
|---|---|---|---|
| through | shine | line | move |

| | | | | | | | | | |
|---|---|---|---|---|---|---|---|---|---|
| d | r | k | o | u | m | g | l | s | i |
| a | s | h | a | d | o | w | n | e | d |
| l | g | t | n | u | v | r | t | c | a |
| i | s | h | i | n | e | k | h | v | r |
| g | a | r | t | h | o | n | r | u | k |
| h | e | g | l | a | s | s | o | r | t |
| t | a | c | h | r | o | k | u | d | l |
| s | l | i | n | e | v | d | g | o | e |
| t | o | w | d | h | i | t | h | e | k |

Shadows

# Use Vocabulary

Finish each sentence.

light    line    shadow

1. Light moves in a ____________________.

2. ____________________ cannot move through some things.

3. A ____________________ is a dark spot.

Draw an object or person that has a shadow.

TARGETING SCIENCE YEAR 2 © PASCAL PRESS ISBN: 9781925726510

Create a picture that has a shadow.

## What You Need

- black, white, and light-coloured cardboard paper
- crayons
- scissors
- glue

## What You Do

1. Draw a picture of an object or person on the white paper.
2. Put the black piece of paper behind the white paper. Hold both pieces of paper together as you cut out your drawing.
3. Draw a sun on the light-coloured cardboard. Glue your drawing onto the paper. Glue the shadow onto the paper.
4. Write a sentence that tells what the picture shows.

# Shadow Walk

Go outside and find shadows.

## What You Need

- a camera

## What You Do

1. On a sunny day, get a camera and go for a walk.
2. Look for shadows outside. When you see a shadow, take a picture of it.
3. After you get home, look through your pictures and answer the questions below.

Draw and write to answer the questions.

1. What shadow did you like best? Draw it.

2. What did it show? ______________________

3. Why was there a shadow? ______________________

______________________



TARGETING SCIENCE YEAR 2 © PASCAL PRESS ISBN: 9781925726510

# Shadow Length Investigaton

For this investigation, you are going to need a piece of chalk and a sunny day.

You will also need someone to trace around your shadow!

## Step 1:

Go outside to a spot in the full sun. Make sure you are not near any trees or buildings, and that the place where you are standing is concrete or a surface you can draw on with chalk.

## Step 2:

Mark a 'x' on the ground. This is where you must stand every time you come outside.

Now get your helper to trace around the shadow your body makes on the ground.

Come out to exactly the same spot every hour, or as often as you can, and get your helper to trace around your shadow each time.

What's happening to the length and position of your shadow?

Can you work out why this is happening?

This might help you to understand.

The sun spins, but stays in the same position in our solar system. The planets move around the sun. As they move around (or orbit the sun), they also spin.

When you are standing on the ground, the Earth is slowly spinning you either towards or away from the sun. The further that particular place on Earth turns from the sun, the longer the shadow will be, because the sun will be at a lower angle to where you are on Earth.

The pictures on the next page will help you to understand this.

# How Are Shadows made?

Have you seen your shadow on a hot sunny day? Ever tried to outrun it? Do you know how a shadow is made?

Shadows happen when an object blocks light. The light source can be natural like the sun, or man-made like a torch.

When you're standing in sunlight, your body blocks the light. Light travels in a straight line, so it cannot bend around you. Your shadow is a space where the light can't reach.

Have you ever noticed that your shadow changes? Sometimes it is long and thin and other times it is short and fat. This depends on the position of the sun at the time. Remember, the sun does not actually move across the sky, rather it is the Earth that is always moving.

At midday, the sun is directly above us, so the light rays are coming from above. There is just a small round shadow at your feet. In the morning, the sun is low in the sky. The light comes from the side, so it makes your shadow longer and skinnier. The same thing happens in the afternoon, but from the other side.

Go outside at different times and look at your shadow. You might be surprised at just how many different shadows you make in one day. You could even take a piece of chalk and get someone to trace it at different times to compare!

TARGETING SCIENCE YEAR 2 © PASCAL PRESS ISBN: 9781925726510

# Describing Sounds

https://clickv.ie/w/7zgx

**Use this QR code to access a video on this topic.**

Sounds can be **loud,** and heard from a very long way away.

Sounds can be **soft**, and very hard to hear.

Do you think these things would make a loud (L) or soft (S) sound?

Write the correct letter in each box.

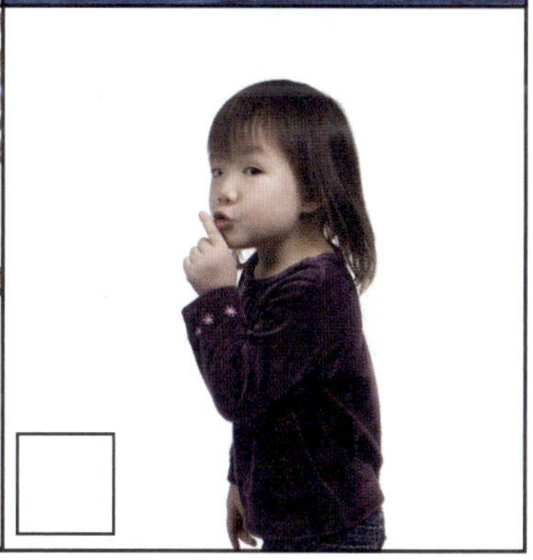

Do you like loud sounds or soft sounds? Can you explain why?

Sound

# High Pitch, Low Pitch or No Sound At All!

Have you ever had someone scream near you? Did it hurt your ears?

Sounds that have a very **high pitch** can be very loud and not very nice. Can you think of some other sounds that are very high pitched?

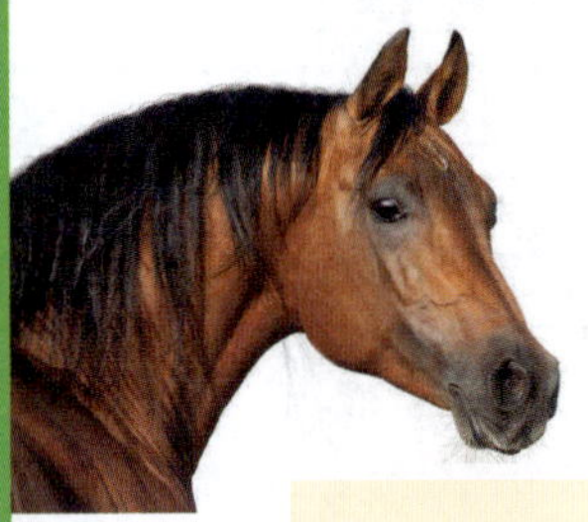

Sounds that are very **low pitched** often make a deep, rumbling sound. Thunder and the nicker of a horse are low pitched sounds. Can you think of any more?

Some things don't make any noise at all, or if they do, you might not be able to hear because it is either too high pitched or too low pitched.

Circle the things makes no sound at all:

TARGETING SCIENCE YEAR 2 © PASCAL PRESS ISBN: 9781925726510

Boom, boom, boom!
That's the sound of a drum.
Where does sound come from?
Sound comes from **vibrations**.

When you hit the top of the drum, it will vibrate, or move over and over very fast.

As the drum vibrates, it makes **sound waves** that move through the air.

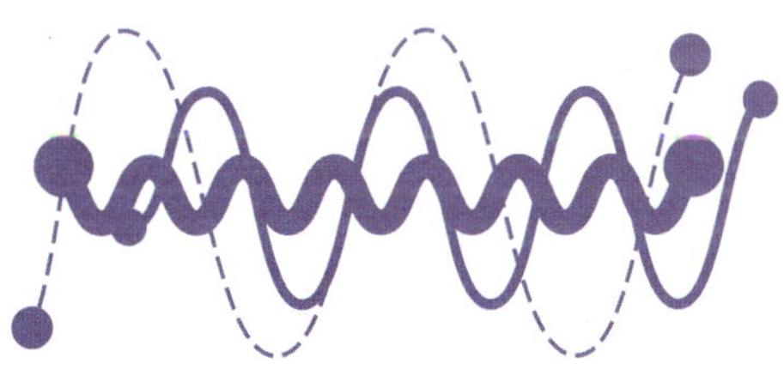

Some sound waves are **musical**.

TARGETING SCIENCE YEAR 2 © PASCAL PRESS ISBN: 9781925726510

# Vibrations Make Sound

We **pluck**
some instruments
to make sound.

We **blow**
some instruments
to make sound.

We **hit**
some instruments
to make sound.

We **shake**
some instruments
to make sound.

TARGETING SCIENCE YEAR 2 © PASCAL PRESS ISBN: 9781925726510

Draw a line to match.

sound waves

musical

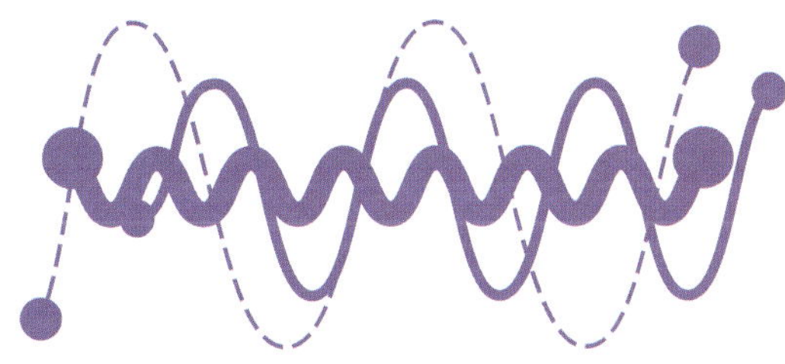

vibrations

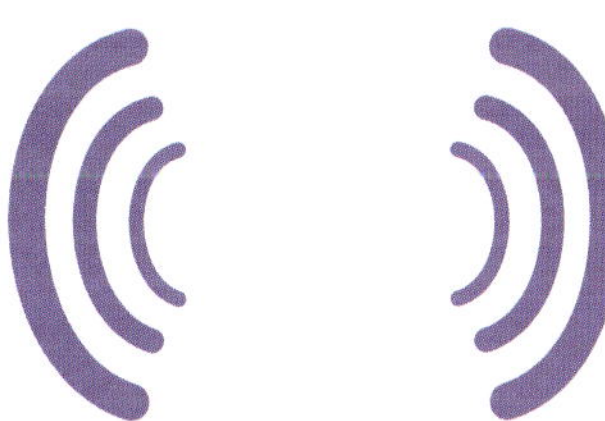

Finish the sentence.

Sound comes from ______________________.



TARGETING SCIENCE YEAR 2 © PASCAL PRESS ISBN: 9781925726510

Draw a line to match.

hit

pluck

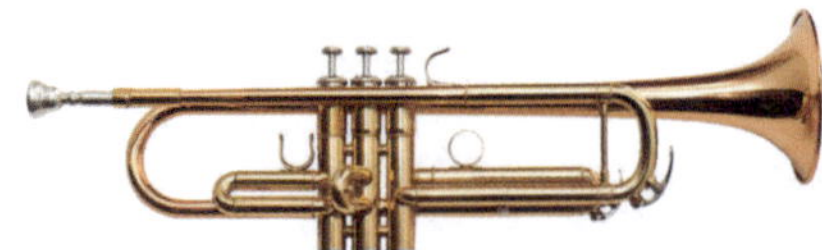

blow

shake

TARGETING SCIENCE YEAR 2 © PASCAL PRESS ISBN: 9781925726510

Read the words in the word box.
Find the hidden words. Circle them.

**Word Box**

| | | | |
|---|---|---|---|
| pluck | hit | vibrations | musical |
| shake | blow | sound | move |

| | | | | | | | | | |
|---|---|---|---|---|---|---|---|---|---|
| v | i | b | r | a | t | i | o | n | s |
| b | l | a | m | v | r | c | h | p | w |
| x | m | u | s | i | c | a | l | l | v |
| h | v | d | c | t | s | o | u | u | a |
| i | a | s | h | a | k | e | a | c | e |
| t | m | o | o | l | c | e | v | k | s |
| u | o | s | e | b | n | n | l | p | o |
| d | v | v | i | b | l | o | w | e | u |
| m | e | c | r | k | e | v | l | t | n |
| b | a | i | o | n | w | i | v | a | d |



TARGETING SCIENCE YEAR 2 © PASCAL PRESS ISBN: 9781925726510

# Use Vocabulary

Finish each sentence.

vibrates move sound

1. When you hit a drum, it ________________.

2. The vibrations make ________________.

3. Sound waves ________________ through the air.

Draw an instrument you **blow** to make sound.



TARGETING SCIENCE YEAR 2 © PASCAL PRESS ISBN: 9781925726510

Use objects from your house to make sounds.

## What You Need

**Set 1**

- a wooden spoon
- a fork
- your hand

**Set 2**

- a glass vase or jar
- a plastic bowl or container
- a metal pan

## What You Do

1. Use each item in Set 1 to tap lightly on an item from Set 2. Notice how the vibrations you create make a different sound for each object.
2. Fill in the chart below to describe each of the sounds you make with the objects.

| | | | |
|---|---|---|---|
| **wooden spoon** | glass | plastic | metal |
| **fork** | glass | plastic | metal |
| **hand** | glass | plastic | metal |

# Making Sounds

Put objects into a cup and shake it up!

## What You Need

- an empty paper cup with a lid
- objects that fit inside the cup: uncooked rice or beans, coins, board game pieces, cotton balls, cord or yarn, etc.

## What You Do

1. Put one group of objects inside the cup. Place the lid on the cup and shake it up. Listen to the sound.
2. Empty the cup and put a different group of objects in it. Repeat until you have experimented with all the objects.
3. Draw which objects made loud or quiet sounds.

| These made a **loud** sound. | These made a **quiet** sound. |
| --- | --- |
| | |



TARGETING SCIENCE YEAR 2 © PASCAL PRESS ISBN: 9781925726510

# Investigations With a Ruler

When something **vibrates**, it makes tiny, very fast, back and forth movements.

When you strum your fingers across the strings of a guitar, they vibrate and move the air around them to make waves of sound.

**Equipment you will need:**

- A flexible plastic ruler
- A desk

**Procedure:**

Take the ruler and place it on the desk with half of the ruler hanging over the edge.

Push down firmly on the end that is on the desk with one hand.

Using your other hand, push down on the loose end of the ruler then quickly let it go so that it flings back.

- Did it make a noise?
- Did the ruler vibrate? Now, move the end of the ruler further out from the edge of the desk. Try flinging the ruler again.
- Did the noise change?
- In what way?

Keep moving the ruler into different positions, and trying the experiment over and over again.

- Can you notice a pattern in the way the ruler vibrates and the way it sounds?
- What are some other things you can think of that vibrate?

# Muffling Sound

Have you ever noticed that sometimes when you make a certain noise in a certain place, it sounds really loud? For example, if you were to sing in the shower.

Other times, when you make the same sound in a different place, the sound is much harder to hear. For example, when you sing out in a large field.

How far sound travels, and how loud it is, has a lot to do with the things around it. Sound waves **bounce off** from some objects, but they are **absorbed**, or sucked in, by others.

Perhaps you could try this!

Get a piece of paper and roll it into a cone. Trying making some different sounds into it, and without it.

What did you notice?

Why do you think this is happening?

An experiment with sound:

For this investigation, you will need something small that can make quite a loud noise all by itself. It might be a noisy toy, a phone or even an alarm clock.

Start the sound, then try wrapping the object in different materials like paper, cardboard, fabric or plastic.

What did you notice about the sound?

We use the word '**muffle**' when we block sounds from escaping. Which material muffled the sound the best in your experiment?

Why do you think this was?

TARGETING SCIENCE YEAR 2 © PASCAL PRESS ISBN: 9781925726510

# Making Music

You don't need a musical instrument to make music!

All kinds of things make wonderful noises when you bang them, strum them, shake them and rattle them.

This man has made his musical instruments out of junk he found at home!

Have you ever tried to make a musical instrument of your own?

Here are some ideas you could try:

# Echo Echo Echo!

Tom was bushwalking with his mother, when they found a deep, dark, cave. When they walked into the cave, Tom's mother called 'Hello' into the darkness.

'Hello, Hello, hello' came back out of the cave.

'Is there someone in there?' said Tom.

'It's an echo', explained Tom's mum.

**Echoes:**

When sound waves travel through the air and hit a hard surface, such as a wall, and then bounce back again, so that you hear the sound again, that's called an **echo**.

Many animals, such as bats and dolphins use **echolocation** to sense objects in their environment and find food.

Bats make a very high-pitched sound from their mouth and nose, and then listen to the echo. This helps them to understand the size, shape and texture of objects in their environment, even when it is really dark.

Dolphins make high clicking sounds and listen to the echo to find food.

Some clever people with vision impairment have learned how to use echoes to find out about what is around them. They might tap a cane, make click clicking sounds with their tongues or snap their fingers, then listen to the echoes to help them find their way. It takes a lot of practise, but it is a wonderful way that they have found to help them in their lives.

TARGETING SCIENCE YEAR 2 © PASCAL PRESS ISBN: 9781925726510

# Musical Instruments - First Nations Perspective

For First Nations people, storytelling and dance is very important.

Music plays a large part in celebrations and ceremonies, as stories, messages, songs and traditions are passed from generation to generation. They sing, dance and tell stories of land, animals and ancestral spirits.

First Nations people make their instruments themselves from materials found in nature. Boomerangs, clapsticks, hollow log drums, seed rattles and the didgeridoo are some of the instruments used to produce the wonderful sounds throughout the rich history of these peoples.

Can you find anything in nature that you could make an instrument from?

What sounds can you make these things produce?

Can you use them to make up a dance to tell a story about something?

# Scientists

Look at the picture. Read the word.
Write the word in the sentence.

**scientist**

A ______________________ studies the world to find out how it works.

**botanist**

A ______________________ is a scientist who studies plants.

**zoologist**

A ______________________ is a scientist who studies animals.

**meteorologist**

A ______________________ is a scientist who studies weather.



TARGETING SCIENCE YEAR 2 © PASCAL PRESS ISBN: 9781925726510

People who study the world to find out how it works are called scientists. In order to learn about the world, scientists ask questions and conduct experiments. They take the information from the experiment and think about the results, or what it tells them. Information that a scientist discovers may be used to solve real-life problems like protecting the environment or finding a cure for a disease.

# Learning About Scientists

Scientists study many different things and work in many different places.

A scientist who studies plants like trees and flowers is a botanist.

A scientist who studies animals is a zoologist.

A scientist who studies the weather is a meteorologist.

One thing all scientists have in common is that they make important discoveries and they help us better understand the world.



TARGETING SCIENCE YEAR 2 © PASCAL PRESS ISBN: 9781925726510

Read the clue. Write the word.

1. A scientist who studies animals.

___ ___ ___ ___ ___ ___ ___ ___ ___
(letter 2 marked: 2nd blank)

2. What a meteorologist studies.

___ ___ ___ ___ ___ ___ ___
(letter 1 marked: 1st blank)

3. Scientists conduct these.

___ ___ ___ ___ ___ ___ ___ ___ ___ ___ ___
(letter 3 marked: 5th blank)

4. Scientists solve these problems.

___ ___ ___ ___ - ___ ___ ___ ___
(letter 4 marked: 5th blank)

5. Scientists make these by doing experiments.

___ ___ ___ ___ ___ ___ ___ ___ ___ ___ ___
(letter 5 marked: 1st blank)

Write the numbered letters to solve the puzzle.

**Science Puzzle**

A scientist helps us better understand

the ___ ___ ___ ___ ___.
1 2 3 4 5

# Scientist's Tools

These are some tools that scientists use to investigate the world. Draw a line to match the picture with the correct definition. Then write the word on the line.

**telescope** **net** **balance** **microscope**

a tool used to gather an animal to study

____________________

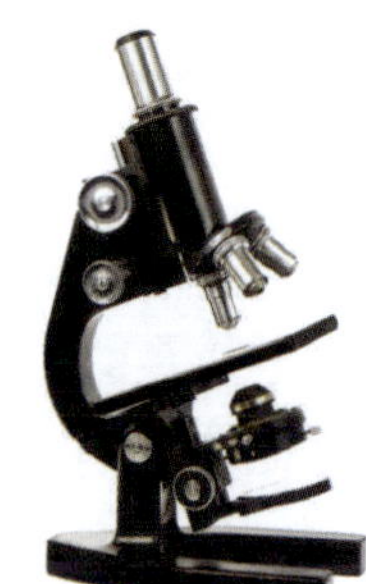

a tool used to make objects in the sky look larger

____________________

a tool that makes small objects look much larger

____________________

a tool used to measure the amount of matter an object contains

____________________

TARGETING SCIENCE YEAR 2 © PASCAL PRESS ISBN: 9781925726510

# Using Science Tools: A Magnifying Glass

Use a magnifying glass to look at things closely and write and draw pictures about what you see!

## What You Need

- a magnifying glass
- an outdoor space

**Information:**
A magnifying glass, or hand lens, is a tool that scientists use to see something more clearly.

## What You Do

1. Use a magnifying glass to observe the things listed below and on the next page. Write and draw about what you saw.
2. Use the example below to help you think like a scientist.

### a leaf

Where I saw it: **under the tree**

What it looks like: **It is orange and has brown veins.**

How it moves/What made it move: **The wind blew the leaf off the tree.**

Science

**an insect**

Where I saw it:

___

What it looks like:

___

How it moves:

___

What it was doing:

___

**Draw It**

**a blade of grass**

Where I saw it:

___

What it looks like:

___

How it moves/What made it move:

___

**Draw It**



TARGETING SCIENCE YEAR 2 © PASCAL PRESS ISBN: 9781925726510

Pretend that you are a scientist. Tell what kind of scientist you are and write about what you are studying. Explain what you want to know and how it may help our world.

# Engineers

Look at the picture. Read the word.
Write the word in the sentence.

**design**

Here is a ____________________,
or plan, of a bridge.

**engineer**

An ____________________ plans the design of something.

**scientist**

An engineer works with a ____________________ to find the best solution.

Sergei Butorin / Shutterstock.com

**builder**

A ____________________ uses the right materials for the job.



TARGETING SCIENCE YEAR 2 © PASCAL PRESS ISBN: 9781925726510

# Engineers

Look all around, and you will see things that people have built. These things are useful.
Before these things can be built, they need to be **designed**, or planned out. A person who designs things is called an **engineer**. An engineer must understand how the parts of something work together. In other words, he or she must know how systems work.

Some bridges have a simple design.▼

▲ This is the Golden Gate Bridge in San Francisco, California. The chief engineer of this bridge faced many problems during the planning. The Golden Gate is one of the longest and tallest bridges in the world, so building it was quite a task!

# Engineers

An engineer asks questions in order to create a good system. Here are some things engineers think about before designing a project:

- How much money can we spend?
- When does it need to be done?
- What materials can we use?
- If it breaks, can we fix it easily?

To answer these questions, an engineer works with other people, such as **scientists**, **builders**, and business people. Together, they figure out the system that works best.





TARGETING SCIENCE YEAR 2 © PASCAL PRESS ISBN: 9781925726510

Read the clue. Write the word.

1. To plan out

____ ____ ____ ____ ____ ____
(letter 2 = 4)

2. A person who designs things

____ ____ ____ ____ ____ ____ ____ ____
(letter 4 = 2)

3. Parts of things working together

____ ____ ____ ____ ____ ____
(letter 6 = 3)

4. A person who knows about science

____ ____ ____ ____ ____ ____ ____ ____ ____
(letter 6 = 1)

Write the numbered letters to solve the puzzle.

**Science Puzzle**

An engineer needs to think about money and ____ ____ ____ ____.
(1 2 3 4)

Engineering

# Good Design

These are some things that engineers design. Draw a line to match the picture with the correct sentence. Then write the word on the line.

**light car bridge fireplace**

It takes you on the road.

____________________

It lets you see in the dark.

____________________

It heats a home.

____________________

It lets people pass over water.

____________________

To find the best design, engineers make models of their projects. Design a strong garage and a weak garage using different materials.

## What You Need

- toothpicks
- mini marshmallows
- craft sticks
- rubber bands

## What You Do

1. Use the materials to make models of two different garages. Make one garage that you think will be strong and will stand by itself. A toy car should fit inside the garage.
2. Make a second garage that you think will be weak and might fall down. A toy car should also fit inside this garage.
3. Watch both garages until the weak one falls over.

Write about your designs on the Garage Record Sheet on the next page.

## Garage Record Sheet

1. What materials did you use for your strong garage?

______________________________________________

______________________________________________

2. What materials did you use for your weak garage?

______________________________________________

______________________________________________

3. My strong garage has... _____ square shapes

_____ rectangle shapes

_____ triangle shapes

4. My weak garage has... _____ square shapes

_____ rectangle shapes

_____ triangle shapes

5. What made the design of the strong garage better than the design of the weak garage?

______________________________________________

______________________________________________

______________________________________________



TARGETING SCIENCE YEAR 2 © PASCAL PRESS ISBN: 9781925726510

Pretend that you are an engineer. Write four questions that you might ask yourself when designing a project.

https://clickv.ie/w/Gygx

Use this QR code to access a video on this topic.

# Physical Changes to Materials

Have you ever tried to change something, without adding anything to it, or taking something away? We call this a **physical change**.

Let's try to make some physical changes to a piece of paper. You will need a few sheets of paper. It doesn't matter if it already has writing on it, as long as you don't want to keep it!

Use this table to write down how many ways you can change the paper. The first few have been done for you to give you an idea.

| Scrunch it in a ball | Twist it | |
|---|---|---|
| | | |
| | | |

Do you think any of these ways of changing the paper would help it to fly across the room?

Use another piece of paper to try to make the best shape for it to fly.

You may have your own way of folding paper to make a plane, but if you don't, you could try folding it this way:

How far could you get your plane to travel?

What is something you could change that might make the plane go further?

Make the change and try it again. Did it work? Why or why not, do you think?

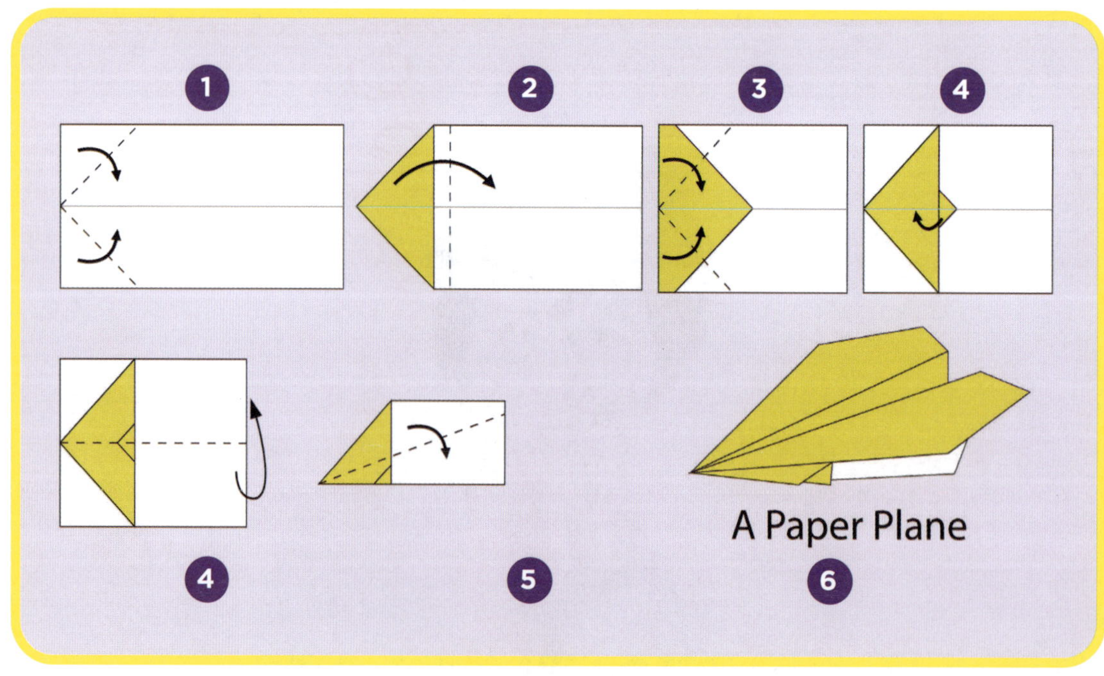




TARGETING SCIENCE YEAR 2 © PASCAL PRESS ISBN: 9781925726510

For this investigation you will need:

- A reel of cotton thread.
- Some weights or heavy things that can be hung off the thread to test its strength. Heavy nuts from a hardware store would work well.
- A few paperclips

**Procedure:**

1. Measure off about 70cm of thread. Tie a nut to the bottom of the thread very firmly. Tie the paper clip to the top of the thread. Make sure it is tied on really firmly. You may need to open the paperclip so that you can feed more nuts over the top of it and down the string.
2. Take the nuts, one at a time, and add them to the thread by pushing them over the paper clip.
3. Each time you add a nut, hold it up by the paperclip to see if it can take the weight.
4. How many nuts can it hold before the thread breaks? ______
5. Now try the same experiment, but this time use two 70 cm threads twisted together.
   Can it hold more nuts? ____________
   Why? ____________
6. Now try it with three 70cm threads.
   Can it hold even more nuts? __________
   Why?
   ______________________________________________

Ship ropes are made up of lots of threads. Why do you think that is?

Where else might people use ropes made of lots of threads?

https://clickv.ie/w/9ygx

Use this QR code to access a video on this topic.

# Physical Change

A **property** is a word that can be used to describe how something looks, feels or acts. Words like strong, heavy or runny are properties that can be used to describe different materials.

Can you find a piece of chalk and a hammer? Or a lump of dirt and a rock to crush it with?

Try this:

Have a look at the piece of chalk closely. What properties does it have? Can you tell someone, or make a list?

Now take the hammer and give it a good hit, to crush it up. Have the properties changed? They should have.... But is it still chalk?

I hope you said 'Yes'!

We can change what some materials look like, but not what they are. This is called a **'Physical Change'**. Below are some ways we can change a piece of paper. Can you write the property words from the bottom to the picture of the way the material has been changed?

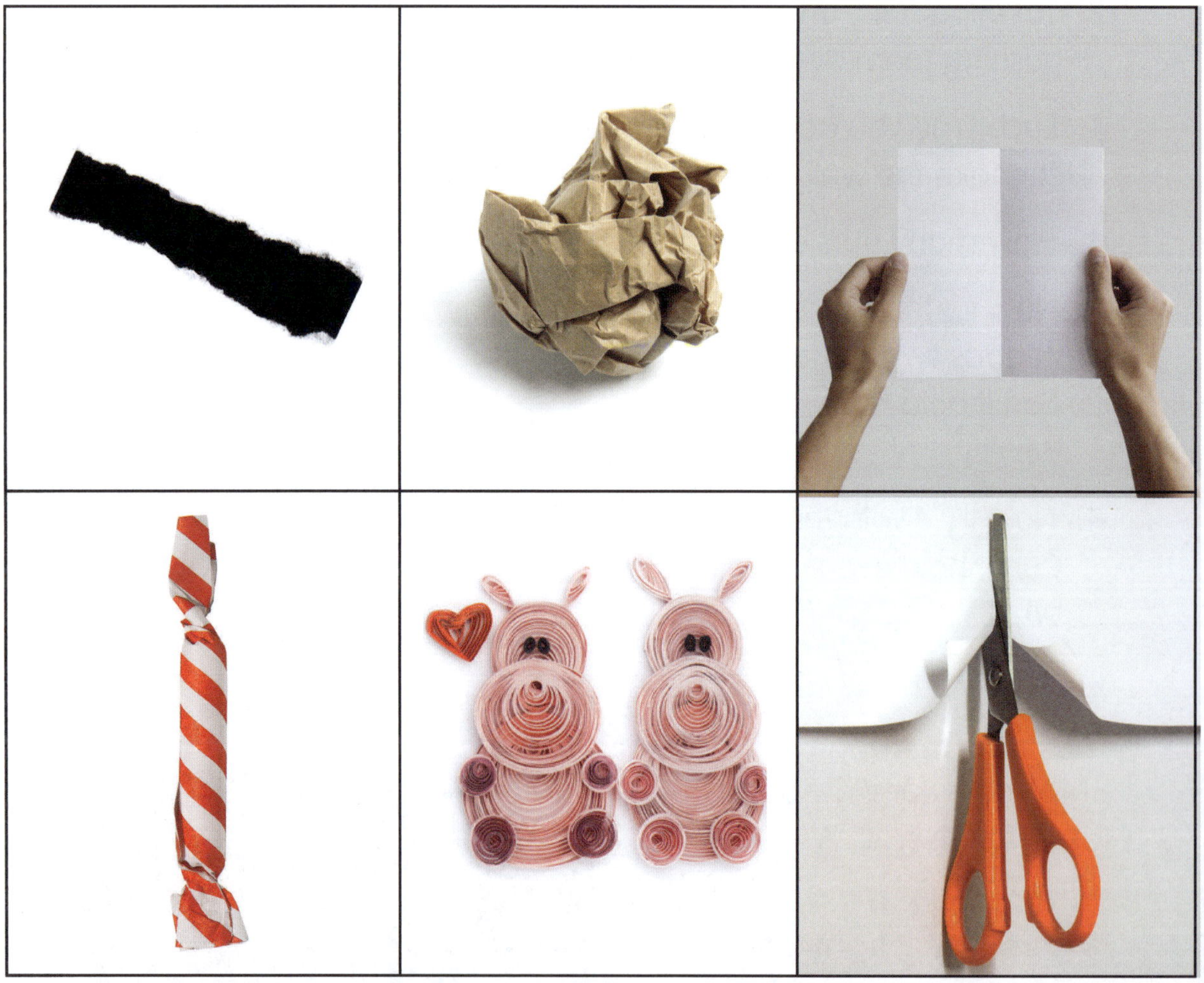

| | | |
|---|---|---|
| | | |
| | | |

| Twist | Cut | Scrunch | Curl | Tear | Fold |
|---|---|---|---|---|---|





TARGETING SCIENCE YEAR 2 © PASCAL PRESS ISBN: 9781925726510

First Nations people understand that many natural materials can be made into very useful things, because of the properties they had.

They use wood to make things like weapons, bowls and shields, and they use the roots, stems and leaves of plants, and skins of animals to make things like shelters, clothing, mats and baskets.

In this picture, a First Nations woman is teaching a young woman how to weave a mat using grass, just by bending it and twisting it.

When it is finished, it will look like this:

Below is a picture of a bowl made by a First Nations Australian. They know the wood is strong, waterproof, light, durable and rigid, so it would be a good material to carve and make a bowl to put food in.

Have you ever made something out of something from nature? In the past, First Nations people had to find everything they needed from the area around them. They were very clever at choosing materials with just the right properties to do the job.

Perhaps you could look around the area you live for some materials you could make into something that has a purpose?

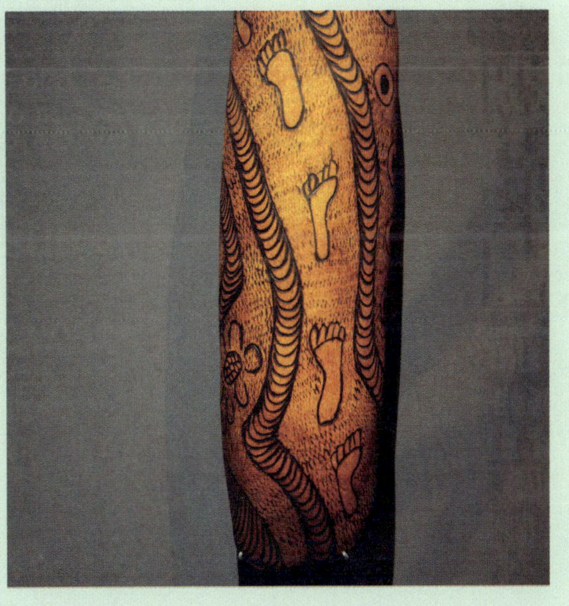

Earth has rocks, soil, and water. These are called earth materials. People, plants, and animals use earth materials to live.

**Rocks** have many different colours, sizes, and shapes. Some animals make their homes in rocks. People use rocks to build homes and buildings.

TARGETING SCIENCE YEAR 2 © PASCAL PRESS ISBN: 9781925726510

Draw a line to match.

rocks

water

soil

Finish the sentence.

Soil, rocks, and water are called ______________________.

Make an X on the one that doesn't belong.

rocks

water

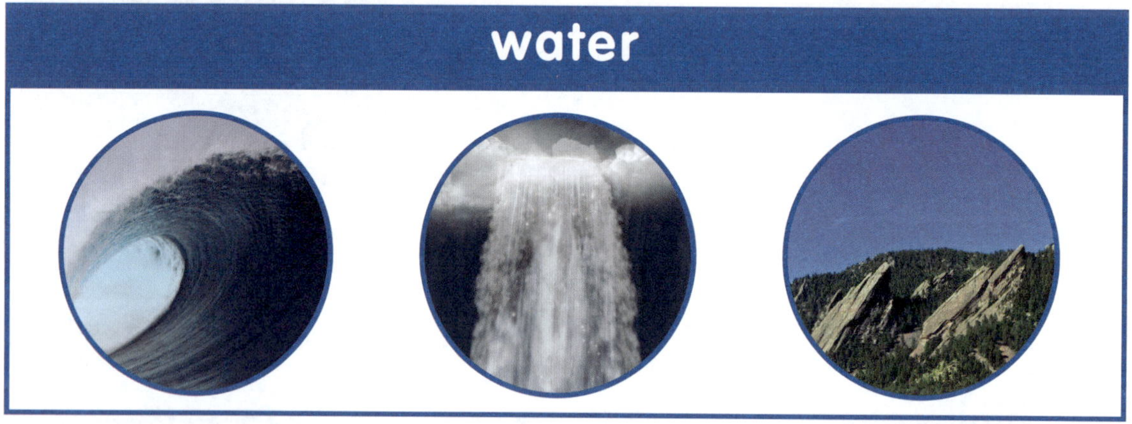

soil

TARGETING SCIENCE YEAR 2 © PASCAL PRESS ISBN: 9781925726510

Read the words in the word box.
Find the hidden words. Circle them.

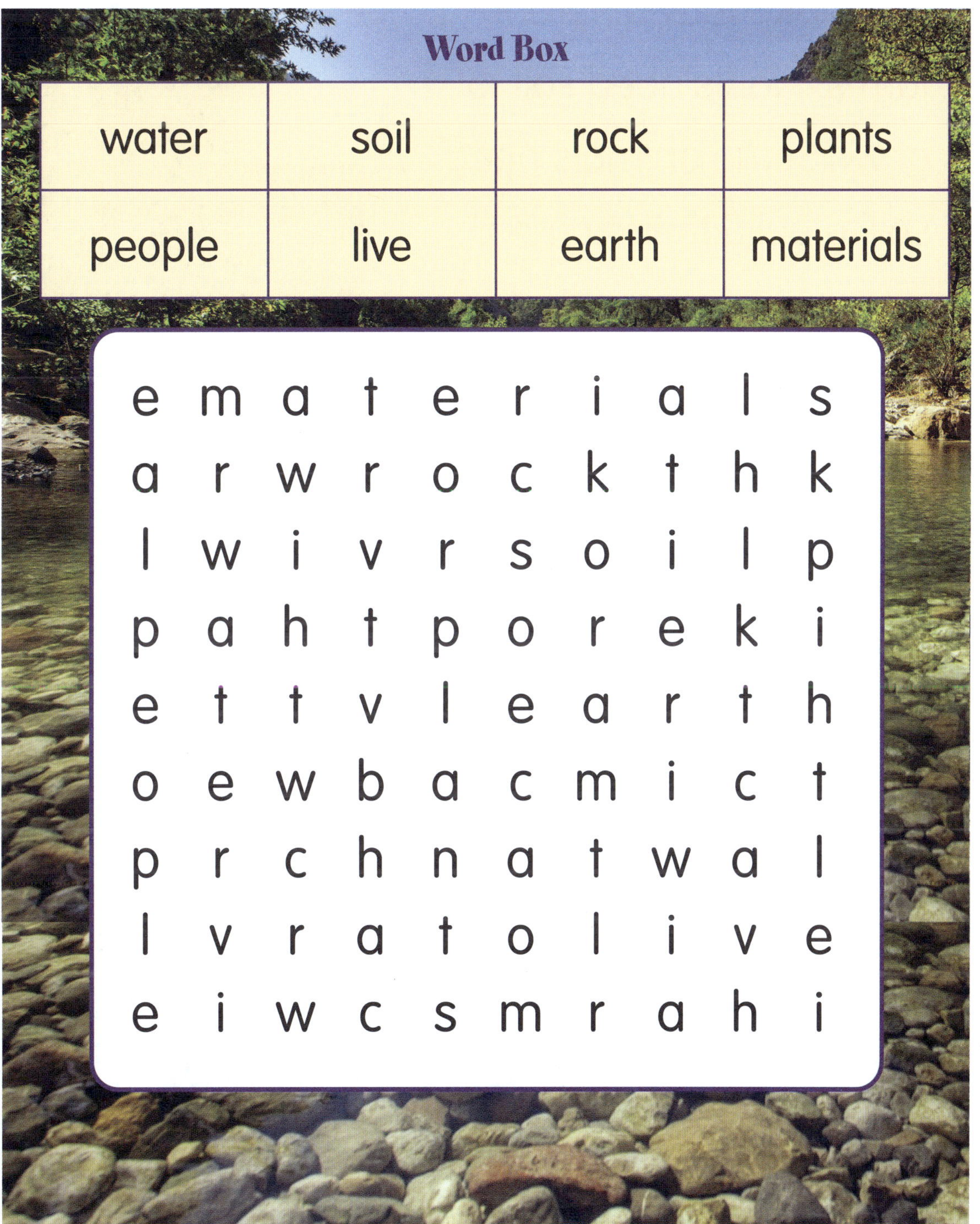

**Word Box**

| | | | |
|---|---|---|---|
| water | soil | rock | plants |
| people | live | earth | materials |

| | | | | | | | | | |
|---|---|---|---|---|---|---|---|---|---|
| e | m | a | t | e | r | i | a | l | s |
| a | r | w | r | o | c | k | t | h | k |
| l | w | i | v | r | s | o | i | l | p |
| p | a | h | t | p | o | r | e | k | i |
| e | t | t | v | l | e | a | r | t | h |
| o | e | w | b | a | c | m | i | c | t |
| p | r | c | h | n | a | t | w | a | l |
| l | v | r | a | t | o | l | i | v | e |
| e | i | w | c | s | m | r | a | h | i |

**Soil** has many different forms. Plants need soil to grow. Some animals need soil to build their homes.

**Water** is found in many places on Earth. People, plants, and animals need water to live. People use water to cook, clean, and do work.

TARGETING SCIENCE YEAR 2 © PASCAL PRESS ISBN: 9781925726510

Finish each sentence.

live rocks materials

1. Earth has soil, ______________________, and water.
2. These are called earth ______________________.
3. People, plants, and animals use earth materials to ______________________.

Draw an earth material you use.

# How I Use Earth Materials

Look around your house, both indoors and outdoors, and find earth materials you use.

How I Use Earth Materials

Draw a picture of how you use earth materials and write a sentence about it.

Answers will vary—Examples:

rocks

We put rocks on the path to our house so we don't walk in mud.

water

I use water to take a bath.

soil

I plant flowers in soil.

## What You Do

1. Think about how you use these earth materials to help you do something.
2. Draw pictures and write about how you use each earth material on the next page.

Draw a picture of how you use earth materials and write a sentence about it.

**rocks**

______________________

______________________

______________________

**water**

______________________

______________________

______________________

**soil**

______________________

______________________

______________________

Trees give us wood. People use wood to build and make things.

We use things made from wood at home, at school, and around town.

TARGETING SCIENCE YEAR 2 © PASCAL PRESS ISBN: 9781925726510

Some things made
from wood are large,
some are small.

Some are used outdoors,
some are used indoors.

Some are for play, and
some are very important.

Even paper is made from wood!
We use wood every day.

# Wood Match

Draw a line to match.

home

school

town

Finish the sentence.

We use wood to ______________________ things.







TARGETING SCIENCE YEAR 2 © PASCAL PRESS ISBN: 9781925726510

Look at the items in each box.
Circle the one that is made from wood.

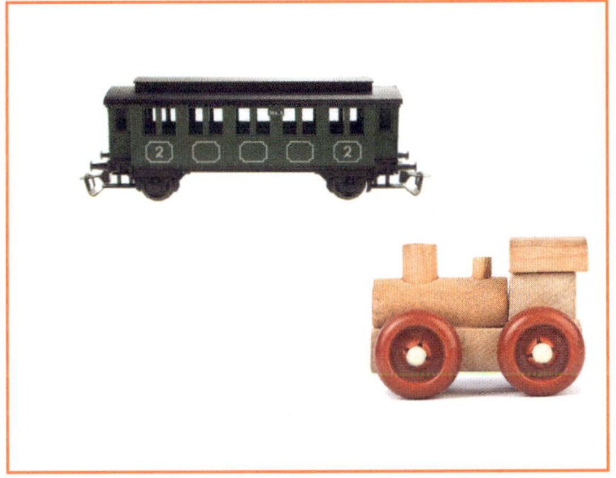

# Wood Words

Read the words in the word box.
Find the hidden words. Circle them.

**Word Box**

| wood | build | trees | house |
|---|---|---|---|
| paper | school | town | table |

| | | | | | | | | | |
|---|---|---|---|---|---|---|---|---|---|
| m | s | c | h | o | o | l | e | e | o |
| r | c | t | r | e | a | c | c | t | w |
| p | a | p | e | r | h | t | b | a | t |
| w | d | w | c | p | a | r | h | b | l |
| b | h | o | u | s | e | e | s | l | c |
| u | n | d | w | o | r | e | n | e | m |
| i | h | t | v | k | b | s | d | c | l |
| l | e | w | o | o | d | a | i | m | r |
| d | r | i | n | c | t | o | w | n | e |


TARGETING SCIENCE YEAR 2 © PASCAL PRESS ISBN: 9781925726510

Finish each sentence.

| town | trees | build |
|---|---|---|

1. ______________________ give us wood.

2. People use wood to ______________________ things.

3. Many things around home, school, and

   ______________________ are made of wood.

Draw something that you use that's made from wood.

# Wood at Home

Look around your house and find things that are made from wood. Use the chart to show what the items are and how they are used.

1. Write the name of each item.

2. Draw a picture of the item.

3. Tell how you use the item.

| Name of Item | What It Looks Like | How We Use It |
| --- | --- | --- |
| pencil | | to write with |
| | | |
| | | |
| | | |


TARGETING SCIENCE YEAR 2 © PASCAL PRESS ISBN: 9781925726510

# I Use Wood to Make Things

Use wooden craft sticks to build something with wood.

## What You Need

- wooden craft sticks
- paper and pencil
- glue
- paint (optional)

## What You Do

1. Think of something you'd like to build with the craft sticks. Some ideas are a pencil or toothbrush holder, an ornament, or a napkin holder.
2. Draw a picture of what you'd like to make.
3. Use craft sticks and glue to make your item.
4. If you use natural coloured sticks, you may want to paint your item.
5. After it dries, use it!

# Odd One Out!

Wood is an amazing material with so many uses. You can cut it, bend it, paint it, sand it, shape it and burn it. You can carve it, pulp it and use it over and over again.

All of the boxes below contain a picture of something that has been made from wood – except for one.

Could you find the item that did not contain wood?

How did you know it wasn't made of wood?




TARGETING SCIENCE YEAR 2 © PASCAL PRESS ISBN: 9781925726510

# Magnets Attract Metals

## Define It!

**attract:** to pull

**iron:** a blue-grey metal that can be made into a magnet

**magnet:** an object that attracts iron

**magnetism:** the force that attracts iron

What makes the notes stick on this refrigerator door? Not glue, but an unseen force called **magnetism**!

Magnetism is the force that makes **magnets** pull, or **attract**, some kinds of metal. The metal refrigerator door is attracted to magnets. A magnet will not attract glass, plastic, wood, or anything else that does not contain metal.

Magnets stick to objects made of metal, but not all metals. The metal **iron**, or a metal that has iron in it, is attracted to a magnet. A magnet won't stick to coins or soft drink cans. They are not made of iron.

1. Draw an **X** on the objects that will **not** be attracted to a magnet.

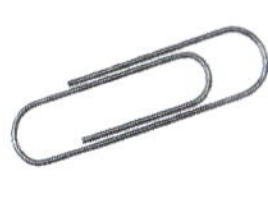

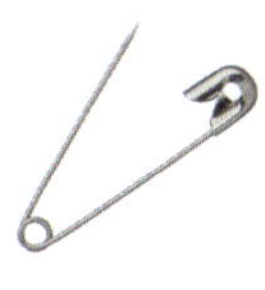

2. Which objects above may have iron in them? How could you test them?

________________________________________

________________________________________

________________________________________

# Magnets Push and Pull

## Define It!

**magnetic field:** the space around a magnet where its force can be found

**pole:** either end of a magnet

**repel:** to push away

All magnets have a **magnetic field** that you cannot see. It's the area around the magnet where a force pulls objects toward the magnet. Magnets have two **poles**. The poles are the parts of the magnet where its force is the strongest. Every magnet has a north pole and a south pole. **N** stands for *north* and **S** stands for *south*.

When two magnets are held with their north and south poles together, the poles attract each other. They pull together. But two poles of the same kind **repel** one another. They push away.

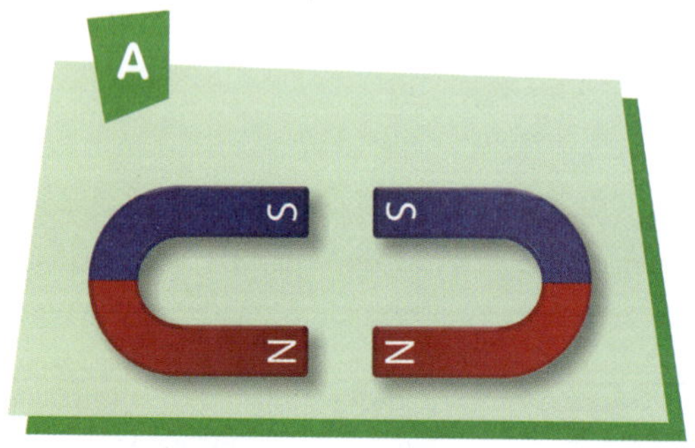

1. Which picture shows magnets that will attract? **A** **B**
2. Which picture shows magnets that will repel? **A** **B**


TARGETING SCIENCE YEAR 2 © PASCAL PRESS ISBN: 9781925726510

## Define It!

**magnetic:** able to attract iron or act like a magnet

**strength:** power

**weak:** lacking strength

Which magnets are stronger—big ones or small ones? You cannot know the strength of a magnet just by its size. The **strength** of a magnet has to do with its magnetic field. A strong **magnetic** field will attract more than a **weak** magnetic field will. The stronger magnetic field will attract things that are farther away or that are heavier.

Look at these two magnets. You cannot see the magnetic force around a magnet. But the lines in these pictures show where the magnetic fields are. The field shown with more lines is stronger. Trace the magnetic fields.

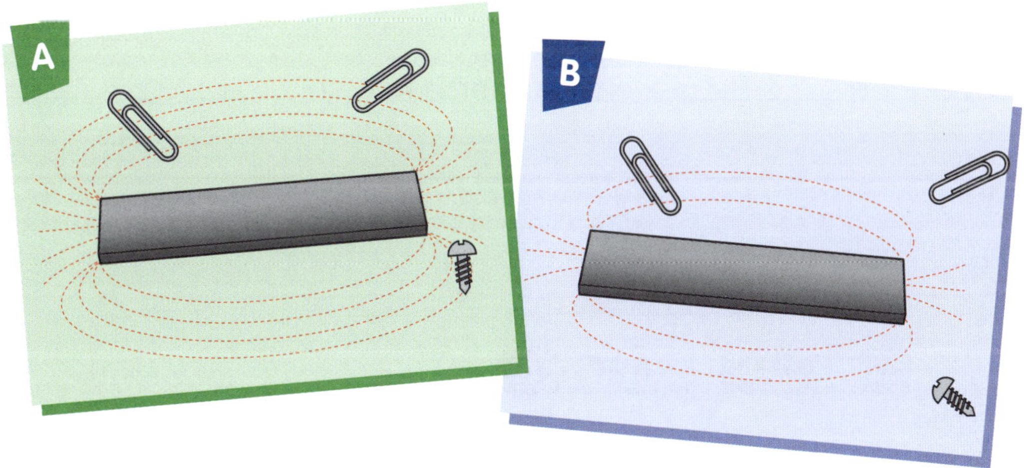


1. Which magnet is stronger? **A** **B**
2. Which magnet will attract the most objects? **A** **B**



TARGETING SCIENCE YEAR 2 © PASCAL PRESS ISBN: 9781925726510

# Looking at Magnets

Magnets **attract** when the north and south poles are put together. Magnets **repel** when two poles that are alike are put together.

Trace the arrows that show the magnetic forces in each picture. Complete the sentence to tell what is happening in each picture and why. Use the words *attract* and *repel*.

1

The magnets __________ each other because ____________________

______________________________________________________________

2

The magnets __________ each other because ____________________

______________________________________________________________

3

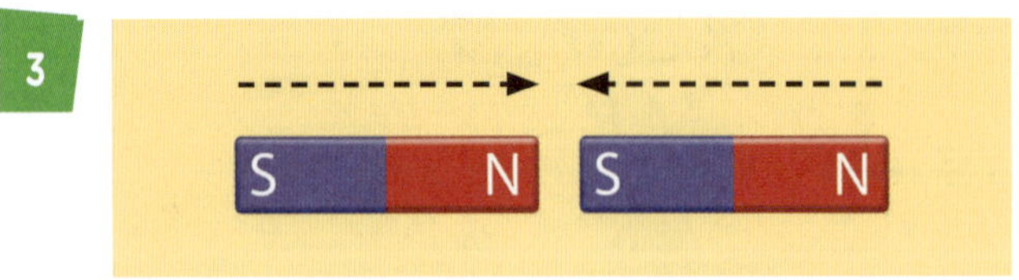


The magnets __________ each other because ____________________

______________________________________________________________

Magnetic Materials


TARGETING SCIENCE YEAR 2 © PASCAL PRESS ISBN: 9781925726510

Read each clue and write the missing word on the lines.

| repel | wood | field | iron | poles |
|---|---|---|---|---|

1. The space around a magnet that attracts metal is its ___ ___ ___ ___ ___.

2. A magnet attracts metal with ___ ___ ___ ___ in it.

3. Another word that means "to push away" is ___ ___ ___ ___ ___.

4. A magnet has two ___ ___ ___ ___ ___: north and south.

5. Magnets do not attract objects made of ___ ___ ___ ___.

Write the letters from the yellow boxes to answer the riddle.

**Science Riddle**

**You cannot see it, but you can see what it does. What is it?**

It is a magnetic ___ ___ ___ ___ ___.

# Magnet Scavenger Hunt

Go on a hunt around your house. First, find a magnet. Then look for items that will stick to the magnet. Which things will **not** stick? Complete this chart. Draw and label six things you found.

| Magnetic | Not Magnetic |
|---|---|
| | |
| | |
| | |

What shape is your magnet? Draw it here.

**Did You Know?**

Magnets come in different sizes and shapes, but they all have a north and a south pole. On a bar magnet or a horseshoe magnet, one end is the north pole and the other end is the south pole. If your magnet is a disc, one side is the north pole. Flip the magnet over, and the other side is the south pole.




TARGETING SCIENCE YEAR 2 © PASCAL PRESS ISBN: 9781925726510

# Make Bottle Cap Magnets

## What You Need

- bottle caps
- patterned paper
- clear packing tape
- gems, beads, buttons
- self-adhesive disc magnets (available in craft stores)
- scissors
- glue

## What You Do

1. Cut a circle from patterned paper to fit inside the bottle cap.
2. Protect the paper circle by placing it between two pieces of clear packing tape. Then trim the circle and glue it to the inside of the bottle cap.
3. You may wish to glue gems, beads, or buttons inside the bottle cap to decorate it.
4. Peel and stick a magnet to the back.
5. Go around the house and discover where your magnetic bottle cap will stick!

**Note:** Do **not** stick your magnet on a computer, monitor, watch, or hearing aid.

# Apply What You Learned

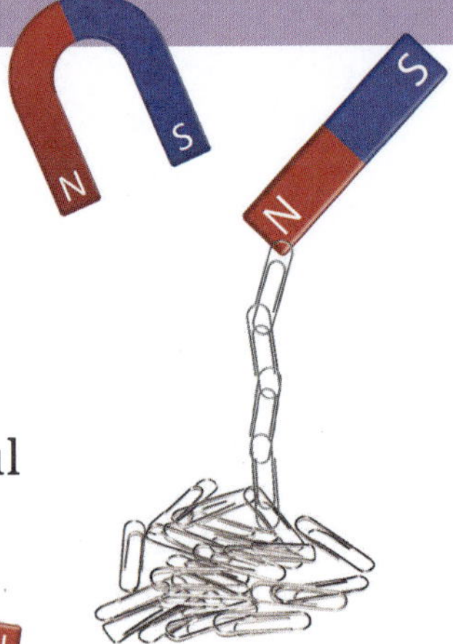


Test the strength of two magnets to find out which magnet is stronger. Use two magnets and a handful of paper clips. Write about your investigation and draw sketches to show what you did.

S N

**Hint**

To test the strength of two magnets, find out how much each magnet can pick up at one time.

1. **Define the Problem:** What do you want to find out?

   ______________________________

2. **Design a Test:** What will you do? ______________________________

   ______________________________

   ______________________________

3. **Record Your Data:** Write what you saw when you did the test.

   ______________________________

   ______________________________

   ______________________________

   ______________________________

   ______________________________

   Draw what you saw.

4. **Write About the Results:** What did you find out?

   ______________________________


TARGETING SCIENCE YEAR 2 © PASCAL PRESS ISBN: 9781925726510

A tool is an item we use to do a job. Tools have different shapes to help us do different jobs.

Straws help us sip.

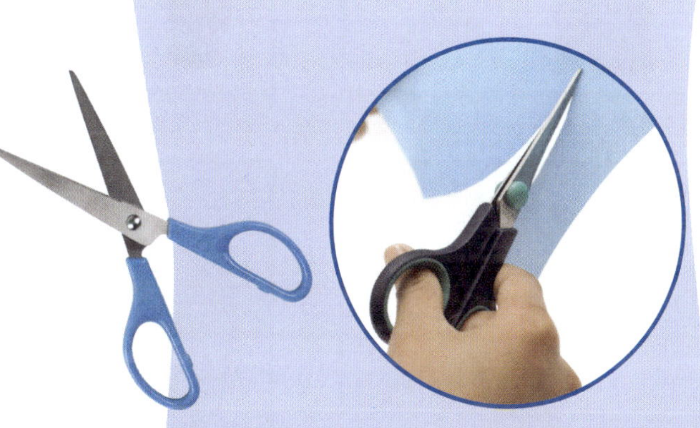

Scissors help us cut.

Ladles help us scoop.

# The Best Tool for the Job

Circle the tool that would be best for each job.

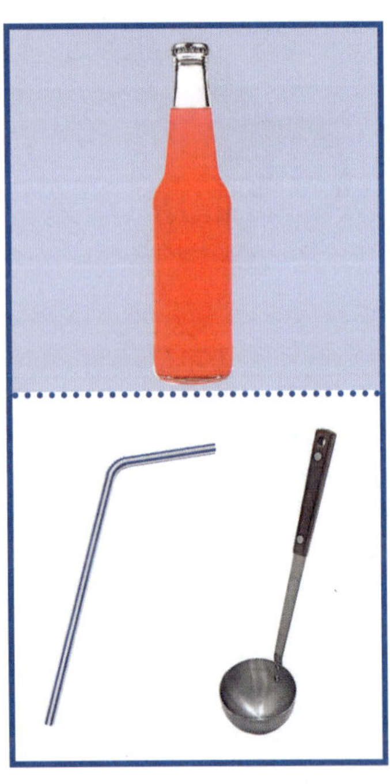

TARGETING SCIENCE YEAR 2 © PASCAL PRESS ISBN: 9781925726510

Finish each sentence.

tool helps shape

1. A ______________________ is an item you use to do a job.

2. Each tool ______________________ you do different things.

3. The ______________________ of each tool helps it do its job.

Draw a picture of a tool that you use every day.

# Paper Funnel Tool

Make a paper funnel and try it out!

## What You Need

- 1 sheet of paper (any kind)
- scissors
- tape

## What You Do

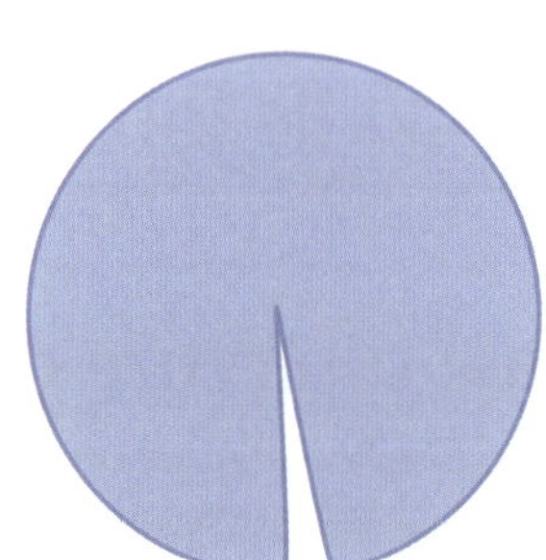

1. Cut a circle shape out of a piece of paper. It does not have to be a perfect circle.
2. From the outside edge, cut a slit to the centre of the circle.

3. Slide one of the cut edges over the other. Continue to slide each of the cut edges opposite one another. Tape the outside edge so that it keeps its shape.

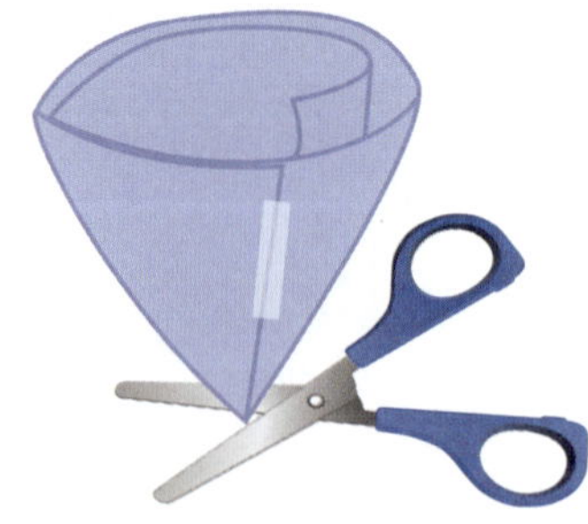

4. Cut the tip to create an opening. The size of the opening depends on what you will be funnelling through it.





TARGETING SCIENCE YEAR 2 © PASCAL PRESS ISBN: 9781925726510

Tell about the paper funnel you made and how you used it.

I made a ______________________________________.

I used it to __________________________________

______________________________________________.

The ____________________ made work easier because

it helped me __________________________________

______________________________________________.

Draw what you poured into your funnel, and the container that you poured it into.

This is how I used my paper funnel.

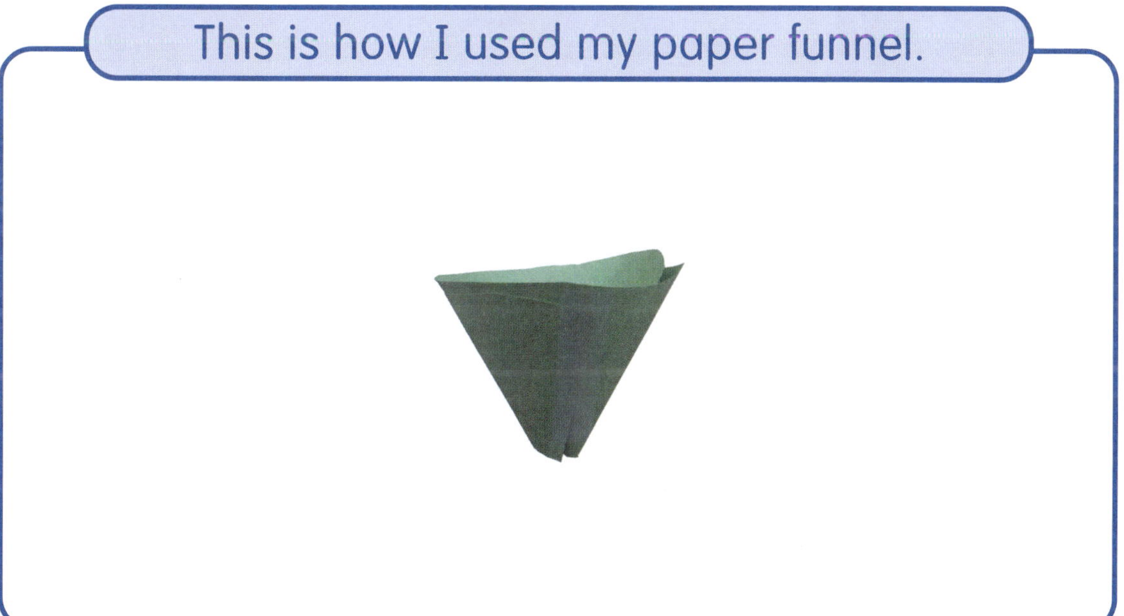



TARGETING SCIENCE YEAR 2 © PASCAL PRESS ISBN: 9781925726510 

# Recycle and Reuse

https://clickv.ie/w/j0gx

Use this QR code to access a video on this topic.

Have you seen this?

It is on rubbish bins.

It is on bottles and cans.

It tells us to recycle or reuse.

We can recycle many things.

glass

plastic

paper


TARGETING SCIENCE YEAR 2 © PASCAL PRESS ISBN: 9781925726510

Some things are not easy to recycle. But you don't have to throw them away. You can reuse them in a different way.

When we recycle or reuse things, we have less garbage. Having less garbage is good for Earth.

# Recycle Match

Draw a line to match.

plastic

glass

paper

recycle

Finish the sentence.

This  means ______________________.




TARGETING SCIENCE YEAR 2 © PASCAL PRESS ISBN: 9781925726510

Draw a line to match. Put the materials in the correct recycle bin!

# Recycling Words

Unscramble the words about recycling.
Write the words in the boxes.

**Word Box**

| recycle | reuse | garbage | cans |
|---|---|---|---|
| earth | glass | paper | plastic |

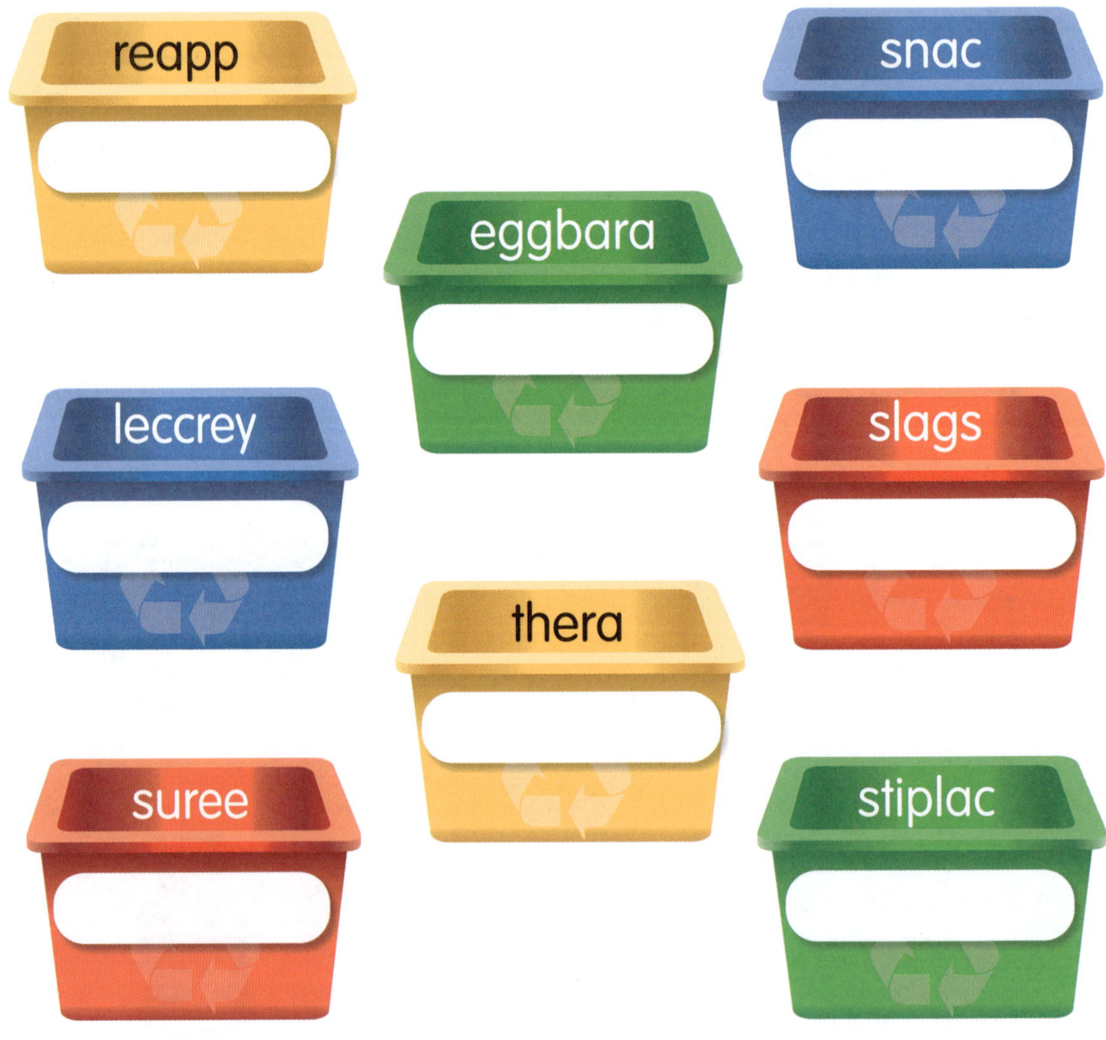





TARGETING SCIENCE YEAR 2 © PASCAL PRESS ISBN: 9781925726510

Finish each sentence.

different recycle garbage

1. We can ____________________ many things.

2. We can reuse things in ____________________ ways.

3. When we recycle or reuse things, we have less ____________________.

Draw things you recycle.

# Birdfeeder

Reuse a milk carton to feed the birds in your backyard!

## What You Need

- 1 litre milk carton
- knife (adult use only)
- hole punch
- paint
- paintbrush
- 5 or 6 thin sticks (twigs or chopsticks)
- glue
- stickers, buttons, or beads
- birdseed

## What You Do

1. Rinse out the milk carton well. Dry it. Set the milk carton on a flat surface.
2. Ask an adult to cut out an opening in one side of the carton as shown.
3. Punch a hole below the opening.



TARGETING SCIENCE YEAR 2 © PASCAL PRESS ISBN: 9781925726510

4. Paint the outside of the carton. After the paint dries, decorate the carton so it looks like it has flowers on it. Use stickers or glue buttons or beads onto it.
5. Ask an adult to cut the thin sticks into halves. Glue the sticks along both topsides of the carton to create a roof.
6. Place a long stick through the hole to make a perch. Glue the end of the stick to the inside back of the carton.
7. Fill the feeder with birdseed and place it outside.
8. Watch the birds use your reused milk carton as a feeder!

# Answer Key

**Page 2**
A round, land; B oceans, clouds, more blue than green; C picture C; D human habitation not evenly spread across planet, by satellite

**Page 7**
1 sun; 2 light; 3 year; 4 heat; puzzle: night

**Page 8**
Answers will vary, but should include information about the sun providing warmth and light.

**Page 9**
The sun faded the parts of the paper that were not covered with stickers.

**Page 11**
The sun is a ball of hot gases in the centre of the solar system. It looks small because it is far away. It gives us light and heat.

**Page 13**
New moon, crescent moon, quarter moon, full moon

**Page 22**
1 full moon; 2 stars; 3 Earth; 4 crescent moon; puzzle: constellation

**Page 24**
crescent moon - slice of moon; full moon - big, round moon; star - ball of hot gases; constellation - pattern of stars; phases - changes in the moon

**Page 25**
The moon gets light from the sun.

**Page 26**
phases, light, shine, sun, moon, star

**Page 27**
1 light; 2 sun; 3 moon

**Page 30**
A season is a time of year when the weather changes significantly in terms of daylight, rainfall and temperature.

**Page 32**
1 seasons; 2 year; 3 weather

**Page 37**
When light cannot pass through something, it makes a shadow.

**Page 40**
1 line; 2 light; 3 shadow

**Page 45**
loud: saw, hammering, mowing, motorbike, rock music

soft: mouse, forest, reading, unamplified harp, waves, cycling, shushing

**Page 46**
feather, grass, butterfly

**Page 49**
Sound comes from vibrations.

**Page 50**
hit - drums; pluck - harp; blow - trumpet; shake - maracas

**Page 52**
1 vibrates; 2 sound; 3 move

**Page 63**
1 zoologist; 2 weather; 3 experiments; 4 real-life; 5 discoveries; puzzle: world

**Page 64**
net - image 4; telescope - image 3; microscope - image 1; scales - object 2

**Page 67**
Answers may vary, but should demonstrate an understanding of making observations of the natural world and thinking about solutions to real-life problems.

**Page 71**
1 design; 2 engineer; 3 system; 4 scientist; puzzle: time

**Page 72**
car - image 2; light - image 1; fireplace - image 4; bridge - image 3

**Page 75**
Answers will vary but may include: what problem am I trying to solve? What materials do I use? How much will it cost? Who will use it? How long will it take? What problems might arise with my solution?

**Page 78**
tear, scrunch, fold, twist, curl, cut

**Page 81**
Soil, rocks and water are called Earth materials.

**Page 82**
rocks - image 1; water - image 3; soil - image 1

**Page 85**
1 rocks; 2 materials; 3 live

**Page 90**
We use wood to make things.

**Page 93**
1 trees; 2 build; 3 town

**Page 96**
The ceramic pot; it has the wrong texture and would not be suitable for purpose if made of wood.

**Page 97**
1 glass, plastic bricks; 2 The metal objects might have iron; you can test that with a magnet.

**Page 98**
1 B; 2 A

Page 99

1 A; 2 A

**Page 100**
1 The magnets repel each other because two north poles are being put together.

2 The magnets repel each other because two south poles are being put together.

3 The magnets attract each other because north and south poles are being put together.

**Page 101**
1 field; 2 iron; 3 repel; 4 poles; 5 wood; puzzle: field

**Page 106**
leaves - rake; height - measuring tape; soup - spoon; drink - straw; pizza - pizza cutter; teeth - toothbrush

**Page 107**
1 tool; 2 helps; 3 shape

**Page 112**
The symbol means the object can be recycled.

**Page 113**
glass: jar, green bottle, brown bottle

paper: carboard roll, magazines, paper cup holder

plastic: lids, plastic bottle, green container, food tray

**Page 114**
paper, cans, garbage, recycle, glass, earth, reuse, plastic

**Page 115**
1 recycle; 2 different; 3 garbage



TARGETING SCIENCE YEAR 2 © PASCAL PRESS ISBN: 9781925726510